S/REF

FEB 1 1 1999

DALY CITY PUBLIC LIBRARY

9043 04645645 0

D0821941

For Reference

Not to be taken from this room

BIOETHICS *for* STUDENTS

HOW DO WE KNOW WHAT'S RIGHT?

General Editor

Steven G. Post

Case Western Reserve University

Carol Donley, Literature Consultant
Brian R. Frutig, Editorial Assistant

Macmillan Reference USA
Elly Dickason, Publisher
Hélène G. Potter, Editor
Anthony Coloneri, Editorial Assistant
Cynthia Crippen, Indexer

Editorial • Design • Production
Kirchoff/Wohlberg, Inc.

The entries in Bioethics for Students *are based, to a large extent, on the* Encyclopedia of Bioethics, *Revised Edition (1995). The Editorial Board of that publication consisted of the following:*

Editor in Chief

Warren T. Reich

Area Editors

Dan E. Beauchamp, State University of New York at Albany
Arthur L. Caplan, University of Pennsylvania
Christine K. Cassel, University of Chicago
James F. Childress, University of Virginia
Allen R. Dyer, East Tennessee State University
John C. Fletcher, University of Virginia
Stanley M. Hauerwas, Duke University
Albert R. Jonsen, University of Washington
Patricia A. King, Georgetown University
Loretta M. Kopelman, East Carolina University
Ruth B. Purtilo, Creighton University
Holmes Rolston III, Colorado State University
Robert M. Veatch, Georgetown University
Donald P. Warwick, Harvard University

Daly City Public Library
Daly City California

BIOETHICS
for STUDENTS

HOW DO WE KNOW WHAT'S RIGHT?

Edited by **Stephen G. Post**

Issues in
**Medicine,
Animal Rights,**
and
the Environment

VOLUME 3

Macmillan Reference USA
New York

R
174.2
BIO
1999
V.3

Copyright © 1999 by Macmillan Reference USA

All rights reserved. No part of this book may be reproduced or transmitted in any form or by any means, electronic or mechanical, including photocopying, recording, or by any information storage and retrieval system, without permission in writing from the Publisher.

Macmillan Reference USA
1633 Broadway
New York, NY 10019

Printed in the United States of America
Printing number
1 2 3 4 5 6 7 8 9 10

Bioethics for students : how do we know what's right? : issues in medicine, animal rights, and the environment/*Library of Congress Cataloging-in-Publication Data* edited by Stephen G. Post
     p. cm.
   Includes bibliographical references and index.
    ISBN 0-02-864936-2 (vol. 1 : alk. paper). -- ISBN 0-02-864937-0 (vol 2 : alk. paper). -- ISBN 0-02-864938-9 (vol. 3 : alk. paper). -- ISBN 0-02-864939-7 (vol. 4 : alk. paper). -- ISBN 0-02-864940-0 (set : alk. paper)
   1. Medical ethics. 2. Bioethics. 3. Science--Moral and ethical aspects. I. Post, Stephen Garrard, 1951–   .
   R724.B4825 1998
   174' .2--dc21
                                    98-29518
                                       CIP

The paper meets the requirements of ANSI-NISO Z39.48-1992 (Permanence of Paper).

CONTENTS

Animals

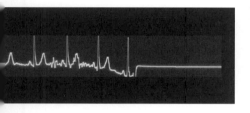

Death and Dying

Environment

Ethics and Law

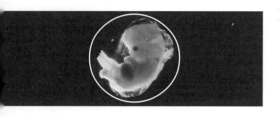

Fertility and Reproduction

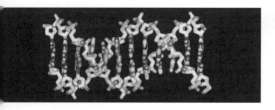

VOLUME 2

Genetics

Health Care

Mental Health

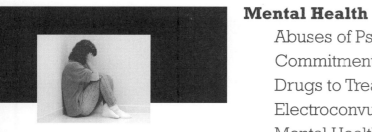

Population

Professional–Patient Issues

Stages of Life

Therapies

VOLUME 4

Therapies (Continued)

Transplants and Other Technical Devices

General Topics

Religious Perspectives

BUDDHISM

meditation spiritual introspection or contemplation of a religious truth, often practiced in order to reach a higher state of understanding or realization

dharma in Hinduism and Buddhism, divine law or comforming to one's duty or nature

enlightenment a higher spiritual state in which a person has moved beyond feelings of desire or suffering

The Dalai Lama is the spiritual leader of Tibetan Buddhism. *Dalai* means "ocean" (of goodness and compassion); *Lama* means "superior." Buddhists throughout the world look to the Dalai Lama as the embodiment of compassion.

According to the story of Buddhism, Siddhartha Gautama, prince of the Sakya clan in India (and the future Buddha), lived in a palace, away from human suffering. At the age of twenty-nine, the prince ventured for the first time beyond the walls of his palace. On his journey, he came upon four individuals—an old person, a sick person, a corpse, and a holy beggar. These chance meetings greatly troubled the prince. The first three individuals represented the suffering that we, as human beings, experience. In the fourth, however, the prince saw a quietness that went beyond suffering. The beggar represented the way the prince would come to understand suffering and, through understanding, go beyond it.

For the next few years, the prince practiced **meditation**. Through meditation, he mastered powers of greater awareness. Eventually, he realized the **dharma**, "truth," and became the Buddha, "The Enlightened One." The experience of **enlightenment** is described as "realizing the undying" and "the discovery of the path to freedom." The truth that Siddhartha Gautama realized is "interdependence" *(pratityasamutpada)*, the core of Buddhist thought and practice. Interdependence means that all living beings and things are related and depend on one another.

The realization of the truth of interdependence freed the Buddha from his ignorance of the three basic features of samsara. Samsara is the realm of change and suffering. The first feature of samsara is that nothing is permanent. All things are constantly changing. The second feature is that nothing lasts because of this constant change. The third feature of samsara is suffering, which results from living in a world that is constantly changing. The Buddha found wisdom *(prajna)* and moral conduct *(sila)* to be the paths to nirvana. Nirvana means perfect contentment, and achieving this state is the goal of the Buddhist life. The Buddha's life and teachings have made his followers sensitive to the needs of others. They have also served as models for Buddhist ethics.

Buddhist Doctrine

On the morning of his enlightenment, Siddhartha Gautama realized the profound idea that all beings and all things are linked and mutually dependent. He understood that no single cause determines how a being or thing is created or destroyed, or how an event happens. Rather, the reality of what we are, the events we experience, and the actions we take are the result of countless related causes and conditions.

doctrine a set of beliefs taught by a particular religious group

karma deeds and their good or evil quality that affect the doer's spiritual condition, in his or her present life or in a next stage of existence

spiritual relating to the nonphysical intelligence or feeling part of a person; having to do with sacred matters or religious values

illusion a false view of reality

A home altar decorated with fruit and flowers honors Buddha during the annual New Year celebration.

Buddhists have based their **doctrines** and practices on two basic ideas central to the truth of interdependence. First, interdependence is a principle of cause and effect. No event has only one cause. Second, interdependence describes the idea that all beings and things are related and dependent on one another.

Karma. As an ethical law, **karma** states that the present condition of a person is determined by his or her past conduct. The future is determined by present decisions and actions. Moral behavior leads to **spiritual** growth. Evil leads to spiritual decay. Molded by his or her past and by the present circumstances, an individual's actions affect other people. Ultimately, these actions also affect the universe.

The Four Noble Truths. The Four Noble Truths were the Buddha's method for achieving spiritual health. They relate directly to karma and to the idea of interdependence. The Four Truths describe the condition of our lives. They also explain suffering and the means by which we can free ourselves from the misery of the world. The Four Noble Truths are:

1. the Noble Truth of Suffering;
2. the Noble Truth of the Cause of Suffering, that is, **illusion** and desire;
3. the Noble Truth of Nirvana, a realm free from suffering; and
4. the Truth of the Noble Eightfold Path, the path to nirvana.

In the First Truth, the Buddha realizes that spiritual suffering, though the most serious, is just one of many illnesses. The Second Truth is that the cause of this suffering comes from illusion and from desiring things that a person cannot have. A person arrives at the Third Truth, nirvana, by way of the Fourth Truth—the Eightfold Noble Path. This path releases the individual from ignorance.

The Eightfold Path means developing
Right View
Right Thought
Right Speech
Right Action
Right Livelihood
Right Effort
Right Mindfulness
Right Meditation.

"Right" refers to the wisdom to discover the Middle Path between indulging in pleasure and engaging in spiritual exercises. Right View is the understanding of interdependence.

Interdependence. The aspect of interdependence describes the cooperative and supportive relationship among all existing creation. All things and beings arise, go on, and disappear in

A gilded statue of Avalokiteshvara, the bodhisattva of compassion, stands in the Jokhang, Tibet's oldest and most sacred temple.

The Buddhist Precepts are rules to live by:

Do not kill.
Do not steal.
Do not engage in sexual misconduct.
Do not lie.
Do not become intoxicated through drugs or alcohol.

Death and Dying:
Euthanasia and Sustaining Life

relation to everything else. We do not simply exist in the world. We help to create the world through the way in which we act and live. In other words, we shape the world as we are shaped by it.

The idea of interdependence imagines a totally connected universe, where all things and all creatures are responsible for each other's well-being. This also means that all men and women can reach the spiritual goal. Both sexes are also expected to assume equal responsibility for each other and the world.

Buddhist Ethics

Buddhist documents often refer to the Buddha as the "Great Physician." Just as a physician cares for the sicknesses of the body, the Buddha cares for spiritual sicknesses. In fact, the Buddha believed that a healthy body helps to develop a healthy spiritual life.

Buddhists are concerned with social and political ethics. They think about issues brought on by medical and technological advances. Some Buddhists feel that traditional ethical codes of conduct are not good enough for dealing with present-day issues. These matters include such questions as brain death, organ transplant, and euthanasia (mercy killing). Buddhists are also trying to find a decision-making process consistent with a right to die.

The meaning of life and death. The question of brain death and the appropriateness of organ transplant have caused great concern. The controversy hinges on the meaning of life and death. Buddhists have traditionally associated life with consciousness and feeling. They believe that animals and plants are feeling creatures. Since feeling is part of life, many Buddhists are reluctant to come out in favor of organ transplants, especially heart transplants. Death of the mind is not death of the person, they feel. The Buddhist definition of death is whole-body failure. Death is caused by cutting off the breathing of a living being.

Prolonging life. The Buddhists' belief in the idea that nothing is permanent also forms the basis of their concern about organ transplant. Since life is not permanent and death is inevitable, and since the spiritual goal of the Buddhist is to go beyond this world, Buddhists believe that life and death should take their natural course. Rather than extending life through heroic measures, Buddhists would spend energy on care of the dying. Those who favor transplants, however, argue that the gift of life is the greatest gift an individual can give. They believe that the human body is only on this earth for a short time and is ultimately meant to be shared.

Doctor-assisted suicide. Buddhists appeal to the idea of interdependence in their approach to ethical problems. The question of doctor-assisted suicide highlights the Buddhist approach to decision making. By taking into account all aspects of suffering, Buddhists try to balance an individual's wish for a gentle death with the doctor's duty to do no harm and with society's desire to preserve life.

Doctrinal limitations. The dilemmas generated by the advances of modern technology have challenged traditional Buddhist notions. The Buddha was aware of the limitations of the Vinaya, the Buddhist book of ethics, and its ability to respond to future problems. The Buddha always stressed that he was a guide, not an authority. As such, he merely outlined a method for determining proper behavior. Should the Vinaya offer no satisfactory course of action, the Buddha asked his followers to make their own decisions. These choices were to be based on wisdom and compassion. Therefore, the Buddha encouraged his followers to use proper conduct and imagination when trying to solve difficult ethical questions.

EASTERN ORTHODOX CHRISTIANITY

theology the study of God

apostolic based on the apostles of Jesus

hierarchy the ruling leaders of religious groups; bishops and ordained clergy

patriarchs the bishops of the Eastern Orthodox churches

doctrine a set of beliefs taught by a particular religious group

sacred dedicated or set apart for the worship of God or a god; holy

The Eastern Orthodox church considers itself identical with the Church established by Jesus Christ and believes itself to be guided by the Holy Spirit. The church continues into the present time united in its history, **theology**, and liturgy with the **apostolic** Church of the first century.

The Orthodox church is organized as a **hierarchy**, with an ordained clergy and bishops. National and ethnic Orthodox churches, such as the Greek Orthodox church, are led by **patriarchs**, and the Ecumenical Patriarch of Constantinople is granted a leadership position of honor. These Christians are united by tradition and **doctrine** rather than by authority. The church's identity is rooted in the experience of the Holy Spirit in all aspects of its life and in a doctrine that is a basis for its ethical teachings.

In the area of bioethics, this theological basis forms a unified source of values for decision making. At the center of these values is the view that life is a gift from God that should be protected, shared, examined, cared for, and fulfilled in God. Life is considered a **sacred** reality.

Doctrine and Ethics

The Eastern Orthodox church understands God to be the Holy Trinity—a God who is a unity of three persons: the Father, the source of the other two divine persons; the Son, forever born of the Father; and the Holy Spirit, forever being sent forth from the Father. God is seen as a community of three divine persons living in eternal love and unity.

This divine reality—God—created everything that exists, visible and invisible. Human beings are created as a combination of body and spirit, as well as in the "image and likeness" of God. "Image" refers to those characteristics that distinguish humankind from the rest of the created world: intelligence, creativity, and the ability to love, make decisions, and understand right from wrong. "Likeness" refers to the potential for such a creature to be like God in some ways, such as in healing and forgiving.

There are fifteen independent Eastern Orthodox churches, each located in a different area. Most reside in countries in the Middle East and Eastern Europe. The Orthodox Church in America is an offshoot of the Russian Orthodox Church. In 1990 it had around 1 million members.

redemption to set free from the consequences of sin

salvation being saved from the power and effects of sin, or from destruction or failure

baptized having undergone baptism, a Christian sacrament; it is a ceremony marked by a ritual use of water and admitting the recipient into the Christian community

divine relating to or proceeding from God or a god

synergy combined action or operation

spiritual relating to the nonphysical intelligence or feeling part of a person; having to do with sacred matters or religious values

natural moral law what is right and wrong according to what humans know about God and creation

material relating to physical rather than spiritual or intellectual things

The work of **redemption** and **salvation** is done by God through the Son, the second person of the Holy Trinity, who took on human nature in the person of Jesus Christ. This saving work, for all humankind, is received by each **baptized** person through faith, and is expressed through the behavior of these people with the help of the Holy Spirit. This cooperation between the human and **divine** while growing into the likeness of God is called **synergy**.

This growth toward God-likeness can be continuous but is never completed in this life. In the Eastern Orthodox worldview, the eternal Kingdom of God provides a sacred reference for everything. The Kingdom is not only to come in the "last days" but is present in our time through Christ's resurrection and the presence of the Holy Spirit. Within this **spiritual** reality, the purpose of human life is to increase union with God, other persons, and creation. This forms the basis for Orthodox Christian ethics and doctrinal teachings.

Among important aspects of these teachings for bioethics are:

1. the supreme value of love for God and neighbor;
2. an understanding that sees human nature as sinful but also capable of providing rules for living based on a **natural moral law**;
3. the close relationship between the **material** and spiritual dimensions of human existence;
4. the capacity for human beings to make moral decisions and act on them; and
5. the idea that the material world and the spiritual world are the means for people to move toward becoming more like God.

Bodily Health

Protecting life as a gift from God gives the health of the body a significant place in Eastern Orthodox ethics. Orthodox Christian ethics call for "a healthy mind and a healthy spirit with a healthy body." The body is neither merely an instrument nor just a dwelling place for the human spirit. It is a necessary part of human existence and requires attention for the sake of the whole human being. Those things that contribute to the well-being of the body should be practiced, and whatever is harmful to the health of the body should be avoided.

immoral not right or moral; sinful

The sanctity of the body. Practices that contribute to the health of the body are ethically required. Adequate nourishment, proper exercise, and other good health habits are fitting and appropriate, while practices that harm the body are considered not only unhealthy but **immoral**. Abuse of the body is morally wrong. Both body and mind are abused through the excessive use of alcohol and narcotics for nonmedical purposes. Orthodox teaching holds that persons who might be attracted to these immoral behaviors should try to overcome their dependence on them as part of their growth toward being like God.

Healing body and soul. Orthodox Christianity teaches that a person who is ill has an ethical duty to fight against the sickness, which, if left untreated, could lead to death. The moral requirement to care for the health of the body allows the use of healing methods that will strengthen the person's health and maintain life. Two ways of healing are used together: spiritual healing and medicine. Spiritual healing is offered in nearly all services of the church, particularly in the **sacrament** of healing, called holy unction.

The church does not see spiritual healing as the only type of healing or as one that competes with scientific medicine. In the fourth century, Saint John Chrysostom, one of the great church fathers, frequently spoke about his need for medical attention and medications. In his letters, he not only speaks of his own use of medicines but also advises others to use them. Saint Basil, a fourth-century church figure, studied medicine and underwent a variety of medical therapies for his illnesses.

sacrament one of several Christian rites (such as baptism) believed to have been established by Jesus as a means or symbol of divine favor

The Protection of Life

Eastern Orthodox thought holds that life is a gift from God, given to creation and to human beings as a great treasure to be preserved and protected. Just as the care for one's health is a moral duty for the individual person, society's concern for public health is also a moral responsibility.

Throughout its history, the church has always valued the protection of life. During the early days of the rise and spread of Christianity, abortion was widely practiced in the Roman Empire. The church, based on its respect for life, condemned this practice in its church law as murder. The church considered abortion to be exceptionally outrageous because of the defenseless and innocent nature of the victim. Most moral teachings are open to change, but in traditional Eastern Orthodox teaching, abortion is nearly always judged to be wrong. Unusual circumstances, such as a pregnancy that threatens the mother's life, might be judged as a condition that allows for an abortion, but such situations are rare.

Modern Medical Technology and Ethics

The development of medical science and technology has raised many new issues. A study of these issues according to the teachings of the church has produced a collection of beliefs that express the church's commitment to the protection of life.

Allocation of medical resources. The Eastern Orthodox church sees providing medical resources to protect human life as a moral obligation. A system that gives the widest distribution of health-care opportunities possible is morally the most responsible,

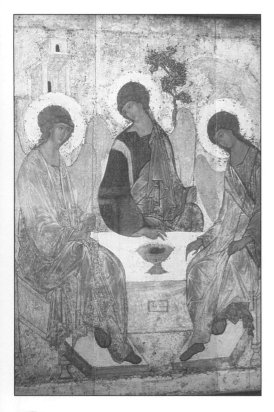

The Trinity of the Old Testament, by Andrea Rublyov (ca.1420–1429).

see also

Professional–Patient Issues:
Professional–Patient Relationship;
Transplants: Organ Transplants;
Stages of Life: Aging and the Aged

since it reflects the church's teaching that all human beings are valuable to God.

Professional–patient relationships. In relationships between medical professionals and patients, the church requires that the medical profession honor patients' rights. The full human dignity of every person receiving medical treatment should be among the most important values of health-care providers. Health-care givers should maintain the patient's privacy, provide the patient with sufficient understanding before consenting to medical procedures, develop warm and friendly personal contacts between the patient and the medical staff, and treat the patient as a total human being rather than as an object of medical procedures.

Human experimentation. Conducting medical experiments on human subjects is permitted and is morally acceptable as long as it values and protects life. Human experimentation can hold the promise of the development of new medical treatments and possible cures for diseases, but the following conditions should be met: (1) The patient should be told of the risks involved; (2) the patient should participate in the experiment freely and without pressure; and (3) the experiment should have the potential to benefit the patient. Increased medical knowledge should always be less important than the welfare of the patient.

Organ transplantation. The protection of life as the primary value must be strictly applied to organ transplantation. Some Eastern Orthodox Christians find organ transplantation wrong because they see it as violating the integrity of the body. As important as this is, it does not outweigh the value of concern for the welfare of one's fellow human beings, especially since organs for transplants are often donated by persons who are motivated by their concern for the protection of life. The sale of organs is considered morally wrong, because it seeks to make a profit from selling human body parts, which the concern for protecting life and human dignity forbids. Yet no one has a moral duty to donate an organ. Health-care professionals must respect the integrity of the potential donor as well as the potential recipient.

The aged. The high value given to the protection of life covers the entire life span of the human person. The Eastern Orthodox church has always had a special respect and appreciation for the aged. Children and relatives should do everything possible to enhance the quality of life for their aging parents and relatives. But when an elderly person is incapacitated, it may become necessary to place the person in a facility for the elderly. Many Eastern Orthodox Christians feel that this is abandoning their moral obligation to their parents. Placing an elderly parent or other relative in a facility for the family's convenience only is morally wrong. Yet even when putting an elderly person in a nursing home is morally acceptable, the children are still obliged to express love, concern, and respect for their parents.

euthanasia mercy killing

There is no consensus on the issue of reproductive technologies. There are those within the Eastern Orthodox church who view some of these methods as ethically acceptable because God has encouraged human procreation.

see also

Fertility and Reproduction: Abortion, Fertility Control

Death. The protection of life is also a concern at the end of life. Death should come when it will, without the interference of humans. God gives us life; God should be allowed to take it away. Proponents of **euthanasia** hold that persons should be allowed and may even be obliged to end their earthly lives when "life is not worth living." The Eastern Orthodox church judges this to be a form of suicide and condemns it. If a person does this to someone else, it is considered murder. Eastern Orthodox Christian ethics consider euthanasia to be morally wrong.

Modern medicine has raised some related issues. Since a person's vital signs can be maintained artificially even after death has occurred, it raises the difficult question of turning off "life-sustaining" machines after brain death is diagnosed. Eastern Orthodox tradition has never insisted on the use of heroic means to sustain life in situations where death is near and there is no further medical help for the patient. It has been Eastern Orthodox practice not only to allow a person to die but also to pray for it when a person is struggling to die.

For the church, a "good death" (the literal meaning of the Greek word *euthanasia*) is one in which the human being accepts death with hope and confidence in God, in communion with him, as a member of his kingdom, and with a conscience that is at peace.

Reproductive technologies. Artificial insemination helps people conceive when they cannot do so through normal sexual intercourse. There are differences of opinion in the Eastern Orthodox church regarding this procedure. A major objection is that it is totally unnatural. But since other "unnatural practices" such as cooking food, wearing clothes, using health aids such as eyeglasses and hearing aids, and performing or undergoing surgery are considered morally acceptable, the "unnatural practice" argument loses strength for some people in the church.

Another argument against artificial insemination is that it separates "baby-making" from "love-making," which is one way the spiritual side and the physical side of marriage are united. The intrusion of a third party in providing the semen interferes with the marital unity of the couple.

If the male sperm and the female egg come from the spouses themselves, and the wife carries the baby to full term, the church's ethical objections to these procedures are less severe. Often, however, fertilized eggs are discarded in the procedure. The majority of Eastern Orthodox consider this a form of abortion. The same pattern of ethical thinking applies to similar procedures, such as in vitro fertilization.

Genetics. Genetic counseling seeks to provide information to a couple before they conceive children so that potentially serious conditions in newborns can be known before a child is conceived. In earlier times, both the church and society tried to reduce the cases of children born with severe problems by restricting mar-

see also

Genetics: Genetic Counseling

riage between persons who were too closely related genetically. Genetic screening today is a more accurate method for achieving the same goal. But once a child is conceived and growing in its mother's womb, the church would not consider the termination of the pregnancy as anything other than abortion. An impaired child is still the image of God with a right to life.

Genetic engineering. There is a conflict of opinion among Orthodox Christian ethicists on the subject of genetic engineering. Some Orthodox ethicists value the potential healing possibilities of genetic engineering. In this case, the treatment of the genetic material of a human being to correct problems is looked at positively, as a form of medical therapy. Yet there is concern that these same techniques may be used in attempts to perfect a race or ethnic group. The potential for such misuse and abuse makes Orthodox Christians approach this issue very cautiously.

The common thread running through all these concerns is the high value the church places on human life as a gift from God. Eastern Orthodox Christianity takes a conservative approach to these issues, seeing them as part of what is held sacred.

HINDUISM

Hinduism is a religious system that has grown and developed from an even older religion based on the Vedas, which are ancient and sacred Hindu writings. This older religion was brought by the Aryans, an early people who invaded India over a period of centuries, from about 2000 to 1000 B.C.E. Like many other religions, Hinduism has its roots in an oral tradition. This oral tradition was put in written form in four groups of Hindu sacred texts during a period from about 1500 to 900 B.C.E. The Hindu medical system known as ayurveda developed from traditions, beliefs, and practices as well as from the Hindu writings. As a religious system, Hinduism has a range of values and codes of conduct.

Vedic religious beliefs gradually changed and developed into Hinduism during the Gupta period of India (300–500). This time period is often regarded as the classical age of Hindu India.

Hindu Worldview

transmigration the passing (of a soul) from one state of existence to another; also, to pass at death from one body or being to another

The Hindu idea of the world includes the important belief of **transmigration**. This belief claims the existence of an innermost self or soul, atman, which is ascribed to all creatures, ranging from the highest god to the lowest insect. Whether the atman is in a great or low being, this innermost self remains unchanged. By taking bodily form, this self becomes involved with matter and the material world. While in the body, the being may act in good ways or in bad ways as it lives out its life. When the body dies, the atman is reborn into another body. There it is prosperous and happy or poor and suffering

karma deeds and their good or evil quality that affect the doer's spiritual condition, in his or her present life or in a next stage of existence

spiritual relating to the nonphysical intelligence or feeling part of a person; having to do with sacred matters or religious values

dharma in Hinduism and Buddhism, divine law or conforming to one's duty or nature

nonviolence the avoidance of violence; the principle of avoiding violence

vegetarianism the practice of eating only a diet of vegetables, fruits, grains, nuts, and sometimes eggs or dairy products

ascetic one who chooses to lead a life that lacks physical comforts in order to develop spiritual strength

Religious Perspectives: Buddhism, Jainism

All Hindus recognize three main virtues: self-control, charity, and compassion. They also accept a second set of virtues: nonviolence, truthfulness, the duty not to steal, purity, and a lack of desire for worldly possessions.

caste one of the hereditary social classes in Hinduism that restrict their members' occupations and association with members of other castes

Brahman a Hindu of the highest caste, or hereditary social class, traditionally assigned to the priesthood

ritual having to do with a traditional ceremony or the established form or words of such a ceremony

according to its moral or immoral behavior in its last body. This explanation of unequal life situations is known as the law of **karma** and the process is called samsara.

The force of karma is an example of an ethical system with rules of right conduct and moral values. The system of **spiritual** truths is called **dharma**. Dharma, a term that is difficult to translate, takes into account cosmic order, sacred law, and religious duty.

Transmigration links all living beings in a single system. Unlike the Judeo-Christian and Islamic religious systems, Hinduism makes no sharp distinction between human and animal. The ideas of karma and samsara promote **nonviolence** and **vegetarianism**. Nonviolence, which is less important in Hinduism than in Jainism and Buddhism, is less strict for laypersons than for **ascetics**. In fact, in Hinduism the idea of nonviolence does not interfere with religious warfare, punishment of criminals, or self-defense.

The process of transmigration is considered painful. In fact, the main search of Hinduism has been to find a ''release'' from the cycle of birth and death that will lead to a state of eternal bliss. For the traditional schools of Hindu philosophy, and for the systems of Buddhism and Jainism that sprang from them, knowledge provides a means of escaping this ever-repeating cycle of birth, death, and rebirth.

As do many other religions, Hinduism has numerous gods. The Hindu gods include one prime being, or God, and many other supernatural beings. Some of these beings follow the will of the higher gods, but others oppose it. Battles between gods and demons, light and darkness, and good and evil are important features of the earliest Hindu literature. These themes also play a major role in Hindu beliefs and practices. Some Hindus look upon the world as a place full of demons. They believe that these demons are normally at war with gods and can be powerful forces in causing misfortune and disease.

Individual Conduct

Within the framework of the three aims of life, called *purusartha*, which are acceptable for the person of high **caste**, are a series of rituals and taboos throughout life. According to Hindus, religious rituals beginning before birth and continuing after death mark the progress of life. Members of the **Brahman** caste are expected to devote a considerable amount of time each day to prayer and **ritual**, and members of other castes are encouraged to imitate them.

Purity and hygiene. The aim of many of these religious acts and taboos is to maintain purity. Many of these practices also keep a standard of hygiene necessary for staying healthy in a tropical climate. Some examples of these religious acts and taboos include taking a daily bath, eating with the right hand and washing the body with the left, never eating cooked food left overnight, and forbidding

Mahatma Gandhi (1869–1948) was the leader of India's nonviolent resistance against British rule. With great courage, Gandhi's movement succeeded in freeing India in 1947. He was killed by an assassin.

any contact with human corpses and animal carcasses. In addition, the bodily fluids of others, such as saliva and mucus, are considered polluting. In fact, Hindus avoid contact with anything contaminated by these fluids, such as used dishes or drinking glasses.

Virtuous principles. In addition to a code of purity practices, Hinduism also emphasizes ethical principles in dealing with people. For example, the idea of nonviolent behavior has often been interpreted as actively benefiting others. Hindu writings and practices encourage honesty, hospitality, and generosity. Rules describing how guests are to be received, fed, and taken care of emphasize kindness as a social value. Hindu texts even encourage students to treat parents, teachers, and guests as they would treat a god.

Hindu Medicine

The complex medical system known as ayurveda, "the science of (living to a ripe old) age," developed in India during the first millennium B.C.E. Although according to karma, demons and gods may play a part in making someone sick, they have a minor role in the Hindu medical books. The job of a doctor practicing ayurveda is to restore the balance of a person's body with medicine, purification, massage, diet, and instructions for an appropriate lifestyle. Experience with a variety of medicines and careful observations of the symptoms, course, and treatment of many diseases enables doctors who practice ayurveda to maintain the respect of a large number of South Asians who continue to seek their help.

Ethics of Medical Practice

The activities of the doctor, or *vaidya*, are closely linked with the three aims of Hindu life that guide appropriate behavior. By relieving suffering and adding to human happiness, a doctor fulfills the first aim by carrying out religious duties. From generous fees from wealthy patients, the doctor fulfills a second aim: riches. The doctor achieves the third aim—pleasure—from building a great reputation as a healer, and from the knowledge that many people whom he or she loves and respects have been cured.

The last two aims are not to be treated lightly. The few famous physicians described in story and tradition were not selfless servants of humanity but very wealthy people. In this way, they resembled successful doctors of modern times. Also, there appears to have been nothing that prevented a doctor from advertising his or her skill. As an example, since Hindu and Buddhist medical traditions were closely linked, a Buddhist text the Mahavagga, provides more biographical information than the Hindu sources about medical practice in the same society. This book refers to the money interests of a famous Hindu doctor. In his youth, this doctor, Jivaka,

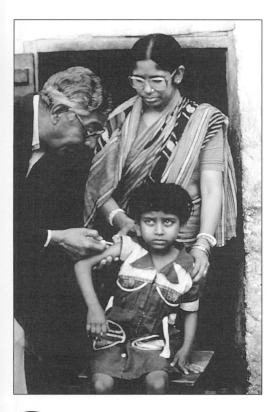

A young Indian girl receives an inoculation in Calcutta, India.

was in search of patients. As he entered an ancient Indian city to earn money for his continuing journey, he walked through the streets asking, "Who is ill here? Who wants to be cured?"

Medical tradition. The Hindu medical tradition is based on a fairly stable theory of health and illness. Despite this it calls for a policy of openness to new ideas about medical treatments. The doctor is advised to be constantly on the lookout for new drugs and treatments. Even after long years of training are completed, the doctor continues to improve his or her knowledge by studying patients and by asking other doctors and even hermits and cowherds about unusual, useful remedies.

Caring for the poor. Providing free medical care to the poor was looked on as part of a king's duty to protect his subjects. From the days of the good Buddhist emperor Asoka in the third century B.C.E., the better rulers of India responded to this responsibility. Medical clinics of one kind or another were established in many Indian cities. There professional doctors gave free services to the poor. Some of these clinics were supported by the government. Others were often paid for by private charity. In South India, especially, hospitals or clinics were often attached to great temples. Doctors themselves may have given money to set up these medical facilities, since they were encouraged to treat rich and poor without charge. But free medical services were more frequent in Buddhist Sri Lanka and Cambodia.

Suicide. The aim of the ascetic to attain release and end the cycle of rebirth has come to be an acceptable reason for suicide in some circumstances. For example, *Sallekhana* is a Jain practice reserved for the elderly poor that involves ritual fasting that ends in death. Its aim is to meet death with the utmost tranquillity. According to Hindu belief, those who undertake a religious suicide in the final stage of life proceed peacefully, living "on water and air, until . . . [the] body sinks to rest."

Questions about these carefully reasoned suicides, usually set aside only for the elderly, were framed in religious rather than medical terms. For this reason, they are different from euthanasia (mercy killing) and assisted suicide in the West. Nevertheless, issues surrounding all these practices are similar, especially the role of terminal illness and great disability.

Although Hindu writings are concerned about ethical questions that either approve of or condemn suicides, the discussion is strikingly different from the present-day debates about doctor-assisted suicides. Suicide in Western countries often raises questions about mental disorder. Concerns for potential victims are focused on prevention and cure of mental disturbances associated with suicidal impulses. But Hindu traditions that consider suicide are concerned with a different set of questions. These questions focus on cultural values. The religious suicides of ascetics and pilgrims, and the suicide of a widow on the funeral pyre of her husband—an

A western woman doing yoga on the beach.

Yoga, a practice that grew out of Hinduism, consists of a variety of techniques to detach the body and the mind from daily life. Classical yoga prescribes solitude and **chastity**. In the Western world, many people practice yoga to attain better health.

chastity abstention from all sexual intercourse; also, abstention from unlawful sexual intercourse

Sanskrit an ancient Indo-Aryan language that is the classical language of India and of Hinduism

act known as sati, after the **Sanskrit** term for the "righteous woman" who undertakes it—were not discussed in medical terms. Modern criticism of sati centers on social, economic, and feminist points of view. It focuses on the sickness not of the victims but of societies that fail to value women, especially widows.

Hindu doctors did not define suicide as an important symptom of mental disorder. However, they did recognize and classify mental diseases according to threatening or disorganized behavior and by disturbing emotional states.

Conclusion

Hindu religious texts, codes of conduct, and writings of ayurveda have recognized and addressed many issues that remain concerns in modern medical practice. The medical books discuss responsibilities of the doctor to society, patients, and colleagues in terms that recognize the professional nature of these interactions, including their social values and political motivation.

Some questions that have become major concerns for medical ethics in the West, such as the status of reasoned suicide, also are looked at in the context of Hindu society, politics, or traditions other than medicine.

Recent developments in technology have placed controversial questions about bioethics and cultural values at the top of the list for a fair and just social policy in South Asia. The ongoing debate that comes from the impact of new technologies in biology needs to be understood in the context of the culture and history that give rise to such hotly debated questions.

ISLAM

monotheistic believing in one God

Along with Judaism and Christianity, Islam is one of the three great **monotheistic** religions in the world. Islam was begun by the prophet Muhammad, who was born around 570.

Historical Development

In the seventh century, Arabia, the region between the Red Sea and the Indian Ocean, was without a strong leader. The city of Mecca had become an important, busy center of trade between Byzantium and nations on the Indian Ocean. When Muhammad was growing up in Mecca, he was aware of the social inequalities and injustices that existed. In Mecca, the tribal society was ruled by a few powerful chiefs.

The Arabs knew about Judaism and Christianity, which were based on a belief in one God. But the Arabs continued to worship their many gods, who were believed to live in holy places in and around Mecca. The most important place of worship in Mecca was

Engraving depicting Muhammad receiving his call to become a prophet in the cave at Hira.

shari'a Islamic law

In addition to most countries in the Middle East, there are large numbers of Muslims in Indonesia, Pakistan, India, and the Philippines. In the United States there are approximately five million followers of Islam.

the Kaaba, a rectangular building that the tribes visited once a year to honor their gods. While they were there, they traded with other tribes who came from all over Arabia.

Before Muhammad, religious practices, beliefs, and values were decided and approved by the tribal chiefs. Their moral code included a belief in "bravery in battle, patience in misfortune, revenge, protection of the weak, resistance of the strong," generosity, and hospitality.

Birth of a prophet. Muhammad was born into the Hashimite clan of the powerful Quraysh tribe in Mecca. His father died before he was born, and his mother died when he was six years old. In keeping with Arab tribal customs, he was brought up by his grandfather. After his grandfather died, he was raised by his uncle, traveling with him to trade in Syria. When Muhammad was twenty-five, he accepted a marriage offer from a woman named Khadija, who was fifteen years older than he was. Muhammad was forty years old when he received his divine call (inspiration) to be a prophet, and Khadija was the first person to become Muslim—a believer in Islam.

Establishing a new religion. The beginning of Islam was a struggle to establish faith in one God instead of many and to create an ethical public order based on divine justice and mercy. The leaders of Mecca resisted Muhammad. They punished Muhammad and his followers, who were mostly poor people not acceptable to the rulers. Because of these terrible persecutions, Muhammad decided to go to Medina, a town in the north, where two Arab tribes who were fighting each other had asked him to decide how they should handle their disagreements. The move to Medina in 622 marks the beginning of the Muslim calendar and the creation of the first Islamic nation. As a statesman, Muhammad put in place a number of changes to organize the tribes into one society on the basis of religious beliefs. One important new belief was that in Islam there is no separation between the religious and everyday activities of human society. Islam insists on the ideal of unity between civil and moral authority under a legal system known as the **shari'a**, which is believed to come directly from God.

After Muhammad

By the time Muhammad died in 632, he had brought the whole of Arabia under the rule of the Medina government. But he left no instructions as to who should take on his religious and political authority after his death. The early Muslim leaders who came after him, called caliphs, used Muhammad's authority to make political and military decisions that led to Muslim rule over lands beyond Arabia. The caliphs believed that the Islamic nation was meant to rule over all other nations, even if those nations did not follow Islam.

The Sunni and Shia split. Differences of opinion on certain important issues arose as soon as Muhammad died. The question of who was to replace Muhammad was one of the major issues that split the community into two groups, the Sunni (also Sunnites) and the Shia (also Shiites). The majority of the people supported Abu Bakr, an elderly friend of Muhammad, in his bid to become caliph. These people gradually became known as the Sunni. Others supported Ali, Muhammad's cousin and son-in-law, to become the imam, a religious and political leader, whom they said had been picked by Muhammad. As the minority, they became known as the Shia.

The dispute was more than just political. A prophet's importance in the community was established in the Islamic holy book, the Qur'an (sometimes spelled Koran), which directs people to obey the prophet. The Qur'an established that Muhammad could use his power to decide how people should live with each other—the public order. This meant that people would have to recognize an authority whose decisions about all areas of Muslim life would be followed.

Two differing views of Islam. The early years of military victories over the Persians and the Byzantines were followed by the civil wars that broke out in 656 under Uthman, the third leader after Muhammad. The conflicts caused by the political and social injustices within the Muslim nation gave rise to two distinct, and in some ways opposing, attitudes among Muslims. Some supported an unquestioning and immediate obedience to almost any Muslim leader who publicly promised to uphold Islamic rule. Others supported activist, radical politics and taught that under certain circumstances it was necessary to remove an unjust leader from power. Gradually, the obedient attitude became connected with the majority of the Sunni Muslims. The activist attitude came to be connected with Shia Muslims.

Shiism upholds the rights of the family of the Prophet to the religious and political leadership of the Muslim community. The name is derived from *shi' at 'Ali*, the Arabic term for the "party" of Ali, cousin of the prophet Muhammad and husband of Muhammad's daughter Fatimah.

▶ The medical writings of the Persian physician Ibn Sina (c. 980–1037), known as Avicenna in the Western world, were accepted well into the seventeenth century.

Muhammad's Ascent to Heaven. The sixteenth-century ink drawing shows the prophet riding a horse and surrounded by angels.

spiritual relating to the nonphysical intelligence or feeling part of a person; having to do with sacred matters or religious values

Fundamental Teachings

The two most important sources of Islamic teachings are the Qur'an, seen by the Muslims as the book of God, and the Sunna, the example of the good conduct of Muhammad as directed by God. The Qur'an is made up of the visions Muhammad received from time to time, from the moment of his call as a prophet in 610 until his death in 632. Muslims believe that the Qur'an was directly spoken to Muhammad by God through the angel Gabriel. Therefore, it is seen as being without error and unchangeable. It has served as the source for ethical and religious beliefs and principles for organizing society. *Sunna* means "trodden path." It has been used to illustrate how the Qur'an has been faithfully followed by devout Muslims, and provides details about ideas and acts traced back to Muhammad's own example. The accounts in the Sunna are called *hadith*. In the ninth century, Muslim scholars developed a complex system for the religious and legal organization of these hadith to support Muslim beliefs and practices.

The First Pillar. The Hadith writings describe the Muslim belief and practice as "the Five Pillars of Islam." The First Pillar is the Shahada, the profession of faith, which states, "There is no supreme being but God, and Muhammad is the messenger of God." Belief in God makes up the completeness of human existence, individually and as a member of society. The Qur'an speaks about God as the being whose presence is felt in everything that exists. Everything that happens is a sign of God. Life is the gift of God, and the body is the trust given to humans to help them to serve God as completely and fully as the wonderful creation of God has made that serving possible.

The Qur'an emphasizes respect for human life by referring to a similar commandment given to others who believe in only one God. The Qur'an says, "We decreed for the Children of Israel that whosoever killeth a human being for other than manslaughter or corruption in the earth, it shall be as if he had killed all humankind, and whoso saveth the life of one, it shall be as if he saved the life of all humankind." This statement gives modern Muslims religious support to justify using medical advances to save lives.

In Islam, the purpose of creation is to allow human beings, created with self-awareness and free will, to accept the responsibility of perfecting themselves. They do this by working with the laws of nature understood by their God-given inborn character and by understanding rules of cause and effect that control their well-being. The Qur'an stresses God's kindness, forgiveness, and mercy. It also stresses that humans should develop *taqwa*, moral and **spiritual** awareness, in carrying out the everyday requirements of life.

The Qur'an's statement that human bodies are restored after death has determined many religious and moral decisions about dead bodies. Dead bodies should be buried respectfully, as soon as

A muezzin calls the faithful to prayer. Muslims are expected to worship five times a day.

possible. Islamic law prohibits damaging the body, so burning the body to ashes also is not allowed.

The Second Pillar. The Second Pillar is daily worship, *salat*, which must be done five times a day. *Salat* must be performed at dawn, midday, afternoon, evening, and night. As Muslims say these very short prayers, they face Mecca and bow. A Muslim may worship anywhere, preferably in a group. Muslims are required to worship as a community on Fridays at midday and on two major holy days—the end of Ramadan and the end of the journey to Mecca.

The group prayer states the believer's religious commitment. Women are not required to participate in group worship, and the Islamic tradition suggests that women worship in the privacy of their homes. But women have always worshiped at certain areas in the Islamic place of worship, the mosque, apart from the men. The Qur'an calls for physical purity of the worshiper, through the performance of cleansing rituals, and a full washing after sexual intercourse or a long illness, prior to beginning worship.

Prayer in Islam is seen as healing. Besides seeking medical treatment, Muslims are supposed to seek healing, especially of mental illnesses, by praying to God. Many illnesses, according to the teachings of Muhammad, are caused by mental conditions like anxiety, sorrow, fear, loneliness, and so on. Because of this, prayer restores the peace and quiet of the soul.

The Third Pillar. The Third Pillar is *zakat*, the required "alms levy." The duty to share what one owns with those less fortunate is stressed throughout the Qur'an. The Muslim definition of a good life includes charity to help widows, travelers, orphans, and the needy.

The Fourth Pillar. The Fourth Pillar is the fast during the month of Ramadan. The Muslim calendar has been in use since the seventh century. Since it is lunar, based on the moon, the month of fasting moves throughout the year over a period of time, because the lunar year is shorter than the year based on the sun.

Ramadan is seen as the holy month during which the Qur'an was given to Muhammad. During the fast, which lasts from dawn to dusk, Muslims are required to refrain from eating, smoking, drinking, sexual intercourse, and acts leading to sensual conduct. The fasting is meant to change the pattern of life for a month, and Muslims have to make necessary changes in their normal schedules of work and study. The end of the month is marked by a festival, Id al-fitr, after which life returns to normal. Ramadan was begun in order to raise Muslims' spiritual and moral self-control. It is also a group experience in which families and friends share both fasting and evening meals in the spirit of giving thanks.

The Fifth Pillar. The Fifth Pillar is the pilgrimage, the hajj. All Muslims are required to travel to Mecca once in their lives, if they have enough money to do so. The rituals of the pilgrimage at Mecca are a group remembrance of the story of Abraham and of lessons to be learned from it.

▶ A nineteenth-century print of Mecca, one of Islam's holy cities, which Muslims are expected to visit at least once in their lives.

The Religious and Ethical Tradition

In the first half of the eighth century, the debates about who should lead, the existence of injustices in the community, and the proper way to fix these problems formed the beginning of the earliest belief system of the group called Mutazilites. Before them, some Muslim thinkers had formed religious arguments—including a principle of God and human responsibility—to defend Islam and the prophet Muhammad when they were challenged by other monotheistic groups.

The Mutazilites. These thinkers tried to show that there was nothing in Islamic beliefs that did not stand up to reason. Their religious system was worked out under five headings. The first heading was the belief in the unity of God, and a rejection of anything suggesting that non-humans have human characteristics. Under the second heading was the idea of the justice of God, and a denial that God's judgment of human beings was unjust. Therefore, humans alone were responsible for all their acts, and would be punished for their evil ones. Third, God's judgment was coming, so it was important to live a righteous life every day. Because of this belief, the Mutazilites did not accept any relaxation in matters of faith. The fourth heading was the idea that a Muslim sinner was in a middle position. Because sinners disobeyed God, they could not be rewarded with Paradise, but because they were Muslim, they could be reformed and would not go to Hell. Finally, the fifth heading commands good and forbids evil in order to have an ethical social order.

The Asharites. Reacting to the Mutazilites' focus on reason, the Asharites limited their religious beliefs to defending the ideas given in the hadith accounts. The Asharites stressed the total will and power of God. They denied that nature and people had any

▶ *Jahangir Preferring a Sufi Sheikh to Kings.* Mughal emperor Jahangir (1569–1627) is shown conversing with a Muslim mystic while King James I of England, the Sultan of Turkey, and a Hindu courtier are made to wait.

important role to play in causing things to happen. They believed that what humans think of as cause and effect is actually caused by God. Hence the idea of free will cannot exist.

Islamic Mysticism

Mysticism is the belief that a person can become perfect, less worldly, and more holy through deep, prayerful thought. Muslim mystics began in response to the growing number of nonreligious groups in the Muslim community. In the early days of the Islamic nation, mystics became an accepted part of Islam. Sufism, as Islamic mysticism came to be known, stressed strict self-judgment and self-discipline in order to reach spiritual and moral perfection. Mystics aimed at obeying the Qur'an in their mind and soul instead of just in the **rituals** of Muslim life. In its early form, Sufism was a form of devotion that involved getting rid of the need to satisfy personal desires in order to devote oneself entirely to God. Sufi mystical practices are done to gain the purity that comes at moments of solitude and deep thought. The Sufi mystics developed intense levels of awareness that could take joyous forms, including a joyous love of God.

ritual having to do with a traditional ceremony or the established form or words of such a ceremony

This part of Sufism brought the mystics into direct conflict with strict, or orthodox, Muslims. Sufism taught that an inner symbolic and spiritual fulfillment of religious duties was as good as participating in actual rituals. Orthodox Muslims saw this teaching as offensive because this community saw strict performance of the requirements of Islamic law as the only correct way to practice Islam.

By the eleventh century, the Sufi masters had come up with a new form of religious belief that brought about the acceptance of Sufism in many parts of the Islamic world. Near the end of the twelfth century, several formal Sufi brotherhoods or orders *(tariqa)* were formed, which women also joined. Each order taught a pattern of prayer and deep thought that used religious practices to bring together a group of beginners under a master. Special forms of breath control and body position went along with words of prayer or sounds to make possible more intense thought. Orthodox Muslims who had not trusted early Sufism were now persuaded to accept the Sufism of the many people who practiced it, and tried to control it.

Islam and Modernity

The modern age brought Islam and Muslims face to face with intellectual as well as political challenges from both within and without. From within, Muslims faced the breakdown of Islamic religious life caused by centuries of decline and the hardening of ideas and beliefs. From without, the rise of the West after the mid-nineteenth century resulted in non-Muslim control of Muslim societies.

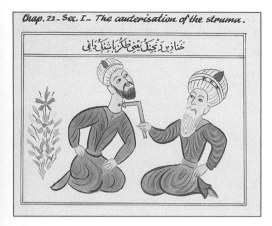

Chap. 22 - Sec. I - The cauterisation of the struma.

خنازیر ز نجیل یغنی طلکر باشنبک داغی

Cauterization of a swollen gland.

Resurgence of fundamentalism. In reaction to these challenges, Islamic fundamentalism has emerged as a movement that holds the Qur'an to be the only source of authority. This movement arose as Muslims became aware of a conflict between the religion that promises earthly as well as holy well-being to its followers, on the one hand, and the historical growth of the Muslim world, which points to the breaking of a holy promise, on the other. Fundamentalist Muslim leaders call for a return to the original teachings of Islam in the Qur'an and the example of Muhammad's life. To regain the power and importance of early Islam, they propose creating the modern nation-state founded on ideals that are based on the practices of the original Muslim community. Muslim brotherhoods throughout the Islamic world have joined forces to put into practice religious changes in a modern society, requiring people to comply strictly with the traditional social and cultural standards.

This resistance to modern nonreligious beliefs and values has posed a major challenge to modernist Muslim leadership. By the late twentieth century, the modernist Muslim leaders had not succeeded in widely convincing the Muslim societies that their alternative to fundamentalist Islam is the only workable solution to modern problems.

Tradition versus medical advances. A case in point is the enormous tension that has arisen along with the technological advances in medicine. Muslim legal scholars face a crisis because Islamic law has halted any modifications within it to adapt to medical progress. Important questions about the role of female doctors and patients in a male-dominated profession; conflict between strict religious observance and medical education; state policy toward family planning; and social and cultural issues that have a bad influence on women's health are among many pressing problems that remain to be officially resolved.

JAINISM

spiritual relating to the nonphysical intelligence or feeling part of a person; having to do with sacred matters or religious values

karma deeds and their good or evil quality that affect the doer's spiritual condition, in his or her present life or in a next stage of existence

The Jaina religion began in India in the sixth century B.C.E., and most of its four million followers live there. According to tradition, the founders of the Jaina faith were not sent by a supreme being, nor were they God appearing in human form. Instead, the founders of Jainism were human beings who through their own efforts reached a high **spiritual** state called *kevala*. *Kevala* is a blissful, all-knowing state of peace and solitude. It is free from the suffering of rebirth into a new life with further obstacles to overcome.

Origin of Jainism

According to Jaina tradition, twenty-four people known as *Tirthankaras* ("ford makers") or *Jina* ("conquerors") crossed the river of rebirth and conquered the influences of negative **karma**

Afterward, they established and spread the Jaina religion. The stories of the Tirthankaras date back to Indian prehistory. But historical records do exist for the last two Tirthankaras, who lived from about 599 to 527 B.C.E. The term *Jaina* means "follower or disciple of the Jina."

Jaina Beliefs

The beliefs and lifestyle of the Jainas are closely linked. There is no creator God in Jainism, which instead is rooted in a unique respect for all life forms. This deep respect for life serves as the basis for a system of ethics based on complete **nonviolence** *(ahimsa)*.

> Jainism takes the principle of nonviolence *(ahimsa)* to include every living being. Animals are never to be injured. To avoid psychological violence, both sides in a dispute are to be respected.

nonviolence the avoidance of violence; the principle of avoiding violence

The Jainas believe that there are two categories of reality: One possesses life *(jiva)*; the other is lifeless *(ajiva)*. Unlike Western definitions of life, which require "metabolism, growth, response to stimulation, and reproduction," the Jainas regard even some objects that seem to be lifeless as possessing life. In fact, the Jainas believe that the universe contains countless life forces grouped into five categories: the first category is earth, water, fire, and air bodies; the other four are microorganisms *(nigoda)*, plants, animals, and humans.

According to Jaina tradition, particles of nonliving matter called karmas stick to beings in the *jiva* categories. This happens when acts of desire, passion, or violence are committed. Though not visible to the naked eye, six colors identify this karma. Black, blue, and gray are connected with sinful or brutish karma. Yellow is linked to less serious offenses. Pink and white indicate that a karmic burden is lessening. By behaving in unethical, passionate, or violent ways, a person adds to the darker, heavier karma. But by following the Jaina code of ethics, one gets rid of the heavy, negative karma and develops the lighter, purer forms. The final goal of Jainism is to break free from all karmic influence. In this joyous state of *kevala*, a person lives forever in energy, awareness, and bliss.

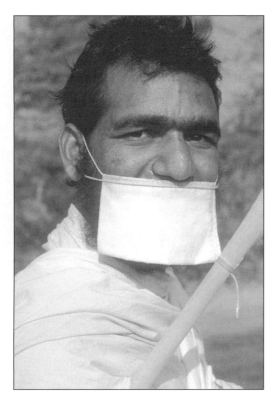

A follower of the Jain sect known as "White-Clad" wears a white robe and mouth covering to keep from inhaling and killing insects (Rajasthan, India).

Jaina Ethics

Jainism's ethics are based on taking vows *(vrata)* to get rid of karma. All Jainas are expected to observe these vows, though the rules for monks and nuns are much stricter. Jaina tradition lists four vows— nonviolence *(ahimsa)*, truthfulness *(satya)*, not stealing *(asteya)*, and not possessing *(aparigraha)*. An early Jaina, Vardhamana Mahavira, is credited with adding a fifth vow: **chastity**, or sexual purity *(brahmacarya)*. Sacred writings such as the Acaranga Sutra serve as Jainism's authoritative sources on religious life.

chastity abstention from all sexual intercourse; also, abstention from unlawful sexual intercourse

From ancient times to the present, Jaina monks and nuns have served as teachers and living symbols of the Jaina tradition. All modern Jainas can be classified as belonging to either the Svetambara ("White Clad") or the Digambara ("Sky Clad") group. In the White Clad group, all monks and nuns wear white robes. In the Sky Clad group, the highest order of monks gives up all possessions, including clothing.

Nonviolence. Jaina monks and nuns wander throughout India, teaching ordinary people about the lives of earlier saints. These religious teachers call for the practice of nonviolence. They discuss such topics as the existence of life forms in the world and the karmic effects of behavior.

Although the vows of the Jaina faith are most keenly observed by monks and nuns, other Jaina followers have developed a culture and way of life based on the practice of nonviolence. Ordinary Jainas generally enter professions that allow them to avoid violent action that would increase the depth and darkness of their karma. Because of this, many Jainas earn their livelihoods through trade and commerce, as long as no animal products and weapons are involved. All Jainas are vegetarians.

Vegetarianism. Although the Jaina religion originally was thought of as showing how to follow a life path of personal freedom and spiritual **enlightenment**, many Jaina practices concerned with avoiding the accumulation of karma have found new importance in modern ethics. For example, **vegetarianism**, animal protection, attitudes toward death, and the belief in tolerance are all relevant to modern life.

Jainas regard not eating meat as a way to be sure that one does not accumulate the negative karma that comes from killing animals. And in modern medical terms also, vegetarianism is seen as purifying the body, lessening the harm to the body associated with eating meat. Jaina eating habits, whose roots are in the ancient belief in nonviolence, are therefore compatible with modern, scientific concerns about strengthening personal health through a low-fat, low-cholesterol diet.

Respect for animals. Respect for animals has long been an important Jaina tradition. Throughout Indian history, Jainas have lobbied for animal protection. They have built shelters and provided food for lost or wounded animals and have successfully campaigned to ban animal sacrifice in most parts of India.

Death and suicide. Jaina tradition regards the death of an older person as both natural and an opportunity to advance spiritually. For many centuries, elderly or sickly Jainas have taken part in a practice known as *sallekhana*, which some modern Jainas refer to as *santhara*. Rather than prolonging death when the process of old age or illness is irreversible, some Jainas receive permission from their religious teacher to fast—to refrain from eating—until death. The

One of the great leaders of the twentieth century, Mahatma Gandhi, followed the Jaina principle of nonviolence throughout his life. The principle also strongly influenced Martin Luther King, Jr.

enlightenment a higher spiritual state in which a person has moved beyond feelings of desire or suffering

vegetarianism the practice of eating only a diet of vegetables, fruits, grains, nuts, and sometimes eggs or dairy products

During the rainy season, Jaina monks spend most of their time in temples so that they won't step on the bugs that swarm in the mud.

Jainas believe that such a fast is not suicide, because it is not done out of despair or hopelessness. They also do not consider it to be mercy killing, because it does not require the assistance of a second person and a violent act. This practice of fasting, associated with a quest for spiritual freedom, embodies the Jaina ideal of meeting and embracing death without fear.

Summary

The Jaina worldview sees the world as a reality filled with life. From the viewpoint of bioethics, Jainism is unique in a number of ways. Its belief in vegetarianism, animal protection, tolerance for different points of view, and its approach to death, in fact, make it a religion rich in ethical content.

JUDAISM

Mishnah a collection of laws orally handed down and compiled, in Hebrew, about 200 C.E., which forms the basic part of the Talmud

Talmud the compendium of laws made up of the Mishnah together with the later Aramaic discussions of them, the Gemara

pluralistic relating to a society in which people of diverse ethnic, racial, religious, and social groups maintain their traditional culture or special interest within a common civilization

covenant a moral obligation of special loyalty

Historically, Jewish life has been rooted in a twofold teaching: the Written Torah and the Oral Torah. The Written Torah is made up of God-given laws found in the first five books of the Hebrew Bible (Genesis, Exodus, Leviticus, Numbers, Deuteronomy). The Oral Torah is made up largely of the laws set down by the rabbis of the talmudic period (first century B.C.E. to the sixth century C.E.) in the **Mishnah** and the **Talmud**, along with the ancient traditions (laws) believed to have been revealed to Moses at Mount Sinai. Regarding many ethical standards, especially those dealing with life and death, Judaism views Torah commandments as binding on all people. This part of Jewish law has been called "Noahide Law," named for the descendants of Noah. Since there is no distinguishing between Jews and non-Jews in questions of life and death, and since almost all medical treatment and so much modern-day discussion of bioethical issues is held within a **pluralistic** society, this part of Jewish law has become the most common basis for most Jewish views on the subject.

Theological and Moral Principles in Jewish Bioethics

A number of religious-moral beliefs are found in Jewish discussions of bioethics. The most common of these beliefs are God as creator, God as **covenant** maker, the sacredness of human life, human goodwill, the authority of the medical profession, and the personal wishes of the patient.

God as creator. The great Jewish religious leaders throughout history all have emphasized that the first belief of Judaism is that God is the creator and Lord of the entire universe, who keeps its everlasting order, its "nature." Because of this, God is considered to be the only possessor of absolute property rights. All creatures

Of the estimated 18 million Jews in the world today, about 9 million live in the United States, about 4 million live in Israel, and about 4 million live in Europe and Russia. The remaining 1 million live in other countries of the Middle East, and to a much lesser extent, in South America, Asia, and Africa.

sanctity sacredness; holiness

martyrdom the acceptance of death as a consequence of holding to one's religious faith

have been given privileges by their divine creator. In keeping with having been created in the image of God, the human creature is given duties as well as the highest privileges. But whatever powers humans have are lawful only when they are seen as having come from God for the sake of God, and not as the property of the individual or the community in any way. "Indeed, all lives are Mine" is from the Hebrew Bible as God's word about this.

God as covenant maker. God is not only the creator of the universe and its everlasting Lord but God is also in a personal relationship with the people of Israel during their long history. This relationship is called the "covenant." Christians and Muslims, who also claim the same God, share in some of this covenantal relationship. This religious belief determines the status of human personhood as being in the image of God. It gives people an ability to have a direct personal relationship with God. This definition of human beings becomes a concern of bioethics because it puts people together with God as they deal with their life situations.

The belief about God as covenant maker is also a major issue in caring for the sick. If the sick have the privilege of making special claims upon those able to care for them, then these privileges and duties are rooted in God's care for his creation. Those with whom God has entered into a covenant must show genuine sympathy to one another. Extending this sympathy to nonmembers of the covenant is viewed as preserving peace and general goodwill.

The sacredness of human life. The term "**sanctity** of human life" does not appear in the well-known Jewish sources. But it is an accurate expression of the belief that one human life has no more value than another, that the blood of one person is not redder than someone else's. The basic principle of the sanctity of each individual human life is expressed in the Mishnah: "Whoever saves even one human life, it is as if he saved an entire world."

However, this does not mean that the value of any human life is limitless. In certain cases Judaism demands **martyrdom**, especially when a person is forced to deny the God of Israel. In addition, at times, priorities are given when only one life in a particular situation can be saved.

The belief in the sanctity of human life can be seen most clearly in cases when there is no reasonable expectation of survival. For instance, if a child is born so defective as to be considered unlikely to survive, he or she is still to be nursed by its mother and not left abandoned to die, as was the case in many ancient cultures.

The treatment of pain is something that may be done as long as it does not bring about the actual death of the patient. One is allowed to pray for the death of the patient in cases where agony is extreme and there is no real hope for recovery. Yet this is always a request for divine action, not a support of people acting in place of God to end someone's life. Even in cases of extreme suffering, the

Maimonides (1135–1204), a Jewish physician who lived in Spain, wrote several medical books, including one on hygiene.

see also

General Topics: AIDS

taking of human life is never to be the purpose of any action. Although a cure cannot always be found, care is always required until the very end of life.

Human goodwill. The duty to care for the sick, and to heal them whenever possible, is taken from two different sets of biblical and rabbinic sources. The difference in the selection of the sources shows two separate approaches to the issue of medical treatment in general.

Maimonides, who was the model rabbi-physician for later generations, saw the duty to care for the sick as a rabbinically ordered act that comes from the general duty of goodwill required by Scripture—"You shall love your neighbor as yourself"—which he summarized as "Everything you want others to do for you, you do." As for the duty actually to save a human life, Maimonides based this directly on the scripturally ordered act: "Do not stand idle by your neighbor's blood"—that is, whoever can save a life and does not do so has broken a commandment.

Protection of the human condition. The human being is always to be cared for, and illnesses are to be cured if possible. The question of the relationship between care and cure is especially crucial today, when the medical means to extend human life is viewed by many as extending the suffering of the terminally ill. Modern Jewish bioethicists struggle with this problem as much as any other group does. There is no complete set of rules on this subject in the Jewish tradition, because the pain experienced at death in the past was seen as being very short. There do not seem to be any rules for dealing with persons in irreversible comas lasting weeks, months, or even years.

Dealing with pain. The most immediate event that medicine treats is pain. A person is immediately aware of pain. All people experience pain at some time. Jewish tradition requires the treatment of unbearable pain in much the same way that it requires the treatment of deadly danger to human life. This can be seen by looking at the laws relating to the Sabbath, which is the most important religious observance in Judaism. Giving relief for pain and protecting human life are specifically permitted by the Sabbath laws.

The public health problem of AIDS involves another challenge to Jewish tradition and its ability to rule in the interest of protecting human persons from all diseases. That challenge comes up when discussing people infected with the AIDS virus by acts that Jewish tradition regards as sinful. Some AIDS sufferers have been infected with the disease through male homosexual acts and intravenous drug use. These acts are forbidden by Scripture and Jewish tradition. One talmudic text minimally prescribes that those who are seen to be "habitual sinners" should be neglected. But the twentieth-century authority Rabbi Abraham Isaiah Karelitz holds that this harsh law no longer applies. Rabbi Karelitz points out that

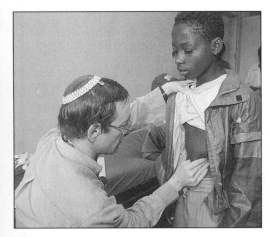

A Jewish doctor examines a young boy at the clinic of the Kigama refugee camp in Tanzania (1996). Jewish teachings demand of doctors that they provide care to those who need it.

the law was originally made to discourage sinners, but in this day and age such a strong law would not be helpful. This opinion is important because it allows for treating AIDS patients with the same concern as those suffering from any illness.

Medical expertise. Jewish tradition has long recognized that a trained medical profession is a necessity in any human society. This can be seen in the Talmud's ruling that no educated Jew should live in a place where there is no physician. So members of the medical profession have special duties and special privileges connected with these duties.

The first duty of medical professionals is to care for whoever needs care. The importance of this duty was noted by Rashi, the great eleventh-century commentator on the Bible and the Talmud. He pointed out the frequent carelessness and arrogance of physicians, and the fact that they often refused to treat the poor. Being poor should not be the reason that a person is denied medical treatment.

Medical ethics. Because medical professionals are involved in an activity commanded by the Torah—and no one is to receive direct payment for obeying a commandment—they are not to be paid directly for their services. Nevertheless, a physician cannot be expected to use up his or her own income on a regular basis to care for patients. If this were the case, only those who were independently wealthy could possibly act as physicians, or as any other necessary community professional. So, medical people can be paid, not for what they do, but for what they might be earning if they were working in another field.

In cases of clear and direct danger to one's own life, Jewish tradition requires the priority of one's own life, whether one is a layperson or a professional. Acts above and beyond the call of duty are considered acts of piety over and above the law. But the real moral problem comes up in situations where there is a danger to those involved in treating the sick. There is a passage in the Talmud that states, "When there is a plague in the city, gather up your legs," which implies that one should save oneself in the face of possible danger. But the sixteenth-century commentator Rabbi Solomon Luria argued that where there is no clear and direct danger to oneself, one should remain in the city during a plague if one is able to save other lives there.

Personal wishes of the patient. Current bioethics emphasizes the personal wishes of those who are ill. Patients can take a more active and responsible role in their own treatment and not simply be the submissive "patients" of medical professionals. Most defenders of patient involvement in medical treatment have looked to the modern principle of **autonomy** for establishing their argument—namely, that human individuals are essentially their own masters. Clearly, the God-centered Jewish tradition does not support patient independence in the strong sense of the term. However,

autonomy self-governing; moral independence

▲ A doctor takes a blood sample from a girl who may be a carrier of Tay-Sachs disease, a genetic condition found in populations of eastern Mediterranean origins.

> Death is better than bitter life or continual sickness.
>
> —*Ecclesiastes 30:17*

it does supply the basis for allowing patients to take an active role for other reasons.

Pain, for example, is to be treated immediately, and the patient is considered the final authority in determining just how much pain he or she can stand, even if that personal determination goes against expert opinion. Unbearable pain is considered worse than death, and to escape it, anything short of direct killing is permitted.

The Stages of Human Life

Judaism is concerned with the human person from conception to death. Especially at the edges of life, where there has been much public debate, Jewish teachings currently have been very much in the forefront.

Abortion. The abortion debate has usually centered on the question of when human personhood begins. Those on the "pro-life" side of the issue argue that human personhood begins at conception, when the egg and sperm join, and abortion is therefore murder. Those on the "pro-choice" side of the issue believe that human personhood begins at birth, and therefore abortion is not murder and should be the option of the individual pregnant woman. Nevertheless, these views are more statements of principle than actual rules. Rules are not directly derived from principles in Jewish law. Instead, principles are formulated to explain rules, balance them with other rules, and guide the ways in which they will be applied to everyday issues.

The rule forbids abortion unless there is a threat to the life or health of the mother. Those who hold that personhood begins at conception see abortion as being the same as murder, although not the kind of murder that would require capital punishment. They would tend to be more cautious in judging what establishes a threat to the life or health of the mother. Yet even they would judge some abortions to be permitted. Those who say that personhood begins at birth still hold that abortion is usually forbidden, because even fetal life has enough rights of its own. It may not be destroyed unless it is a threat to the mother's life or health.

Therefore, Jewish authorities, however they may view the actual beginnings of human personhood in principle, all regard abortion as generally forbidden, and as permitted only under specific conditions.

Definition of death. The question of precisely when human life ends is an issue of much current debate among Jewish bioethicists. Some of them have insisted that the usual guidelines for determining death must be exactly interpreted: the end of spontaneous reflexes, heartbeat, and breath. Yet others have argued that "brain death" can be a good reason for taking a patient off a respirator, because the person's breathing in this case is not being done by the patient but by a machine. In fact, not doing this might violate Jewish law by going against one of its tenets of not leaving the dead unburied. Yet the purpose behind this new thinking is that the interpreters of Jewish law must accept growing medical agreement on any major issue if their rulings are to be taken seriously in the general society, where the most pious Jews receive their medical treatment.

Related Literature

Bernard Malamud's story "The Jewbird" (1961) portrays an ordinary Jewish family who, while having supper one summer night, gets an unexpected visit from a talking bird. The bedraggled bird, much like an unwanted, dependent elderly relative, moves in with the family, asking for a special diet of herring, and constantly irritating the father but befriending and helping the young son. The Jewbird's stay with the family coincides with the illness of the grandmother. On the day she dies, the father throws the Jewbird out. The story may be read as a metaphor for how a family deals with the demanding illnesses of an elderly relative.

NATIVE AMERICAN RELIGIONS

Before the arrival of the European settlers in the fifteenth century, there were over 2,000 different Native American groups on the continent of North America. These ethnic groups have lived through many invasions, widespread diseases, efforts to destroy their culture, and misuse of their belief systems. These groups are also known as American Indians or Amerindians.

The Lifeway

"There is no death. Only a change of world."

—Chief Seattle (1786–1866)

The word "lifeway" means the road of life as native people see it. Such a viewpoint is similar to the idea of a "worldview," a distinct way of thinking about the universe and of judging people's actions in terms of it. Vine Deloria, Jr., a Dakota Sioux lawyer and professor of history, describes an Indian ethical view of the world by saying that in the moral world, all actions, events, and things are related. It does not matter what kind of life a being has, because that being always has the responsibility to be a part of the ongoing creation of reality. This American Indian ethical view sees all life forms as having purpose, being related, and creating together the world they live in. From these views flows a religious system that gives rise to a moral imagination in which the holy is near, within the earth, and can be found in one's thoughts about things that happen naturally.

Black Elk (1863–1950) was an Oglala Sioux religious leader. Black Elk's religious visions, which promised a day of total victory, encouraged the Sioux to fight against the invading whites.

ritual having to do with a traditional ceremony or the established form or words of such a ceremony

vision something seen in a dream or trance that reveals a belief, revelation, or truth

spirit the nonphysical intelligent and feeling part of a person; also, a supernatural being or essence

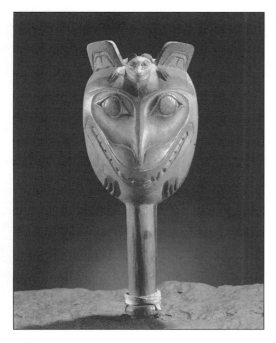

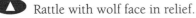 Rattle with wolf face in relief.

All life in a person's environment is dependent on all other life and participates in the act of creation. This can be seen in the changing of the seasons. The term "bioregion" can be used for the Native American respect for all life forms in one's local area. Indians traditionally have understood their local bioregion as filled with moral purpose, wholly interrelated, and alive.

Universal Interdependence

Moral actions in Native American lifeways are acts that are at peace with a holy power believed to fill the world. They happen most directly in the local bioregion. Any one person, through his or her group, is a part of the larger peace. Someone who has committed a crime does not become an outcast because of one bad act. Rather, the person who is out of balance must be brought back, if possible, into the group by **ritual** treatment with that holy power that fills the universe.

Native peoples in North America have words such as *Wakan Tanka* (Lakota), *Kitche Manitou* (Anishinabe), or *Akbatatdia* (Apsaalooke), which show an understanding of the mystery and fullness of the universal power that exists. Holy power and the native words used for it do not show a god that is like a parent; these words stress the web of creative relationships in the religious world and the world of nature. This universal power can be seen in many different **visions**, or **spirits**, that have to do with being close to the power. They may change form in certain holy places, but especially in the local bioregion. Native American lifeways may be described as being "ethically natural." Moral choices come from the desire of individuals and groups to grow within the limits of nature as understood by the people.

Synthetic Ethics

In Native American traditions, ethics and religion are very closely linked. This ethical wholeness is reflected in the practices related to both the rituals and everyday life of native North American peoples. "Synthetic ethics" describes the Native American effort to bring people into the most direct and deep contact with resources for thought and for food. These include the bioregion, the animals hunted, the human community, the seasons, and the holy spiritual world.

Synthetic ethics means the smooth whole of the Native American world, in which people's personal actions reflect the values in their myths, and whose ritual actions show human relationships with the local bioregion. These ethical relationships come from moral examples portrayed in the myth stories. Ideas in the Native American creation stories, such as the living earth and animals with purpose, make one see all experiences as ethical experiences—the seasons,

A Native American carver at work on a totem pole. The pole is a symbol of the bioregion and at the center of all village rituals.

the hunt, or the eating of local foods at their harvest time. The American Indian moral imagination comes from ideas that are believed to rule over personal and group life as well as the bioregion and the universe. This worldview meshes the ethical beliefs needed to deal with events in day-to-day life with ethical beliefs about the world drawn from the peaceful rhythms of nature. The words "lifeway," "synthetic ethics," and "bioethics" mean the total of a good life that is lived in thoughtful relationship with the environment.

Different native peoples have their own words for such ideas as synthetic ethics and lifeway. For example, Winona LaDuke writes:

The ethical code of my own Anishinabeg people of the White Earth Reservation in northern Minnesota keeps groups and individuals in line with natural law. *"Minobimaatisiiwin"*—it means both the "good life" and "constant rebirth"—is central to our value system. In *minobimaatisiiwin*, we honor women as the givers of lives, we honor our *Chi Anishinabeg*, our old people and those who lived before us who hold the knowledge. We honor our children as the continuation of our people, and we honor ourselves as a part of creation. One of the most important things in *minobimaatisiiwin* is living in the same place for a long time, a deep understanding of the relationship between humans and the ecosystem and of the need to keep up this balance.

It is possible to find similar statements by elders from native groups in North America that show the relationship in their lifeway between social justice and the environment.

Land and the Human Presence

The Winter Dance among the Okanagan/Salish/Colville peoples of Washington State exemplifies the special understanding of the relationships between land, lifeway, and synthetic ethics. The Winter Dance shows a developed lifeway where taking part in the ceremony is believed to change people and bioregions. The Salish understand that relationships laid down during the ceremony deepen as a person grows in the ethical path.

The Winter Dance. The yearly calendar of rituals begins with the Winter Dance. Rituals that take place during the calendar year include individual and group activities, such as sweat-lodge ceremonies, vision seeking, stick gambling, healing ceremonies, and first fruits and harvest festivals for deer, salmon, and root crops. The major ceremony, which brings together all the food and healing rituals, is the Winter Dance. This dance is a complex ritual of renewal. It is called by individual hosts from late December through February. A shorter form of the ceremony can be done at any time for someone in need. The center pole is the most important symbol

cosmogony a theory about the creation or origin of the world or universe

spiritual relating to the nonphysical intelligence or feeling part of a person; having to do with sacred matters or religious values

Religious **syncretism** involves the combining of elements of different religions to form a new religion. The Prophet Dance religion emerged among Native Americans in the Southwest during the 1830s. This religion included a traditional Indian round dance, in which participants could receive revelations from the spiritual world while in trance states, combined with certain elements from Christian missionaries such as the symbol of the cross and worship every seven days.

syncretism (or **syncretic** religion) the combination of different forms of belief or practice

of the bioregion. Just by having a ceremony to place it in the middle of the dance house, the healing and changing powers of the Winter Dance can be called forth.

The Winter Dance ceremony is centered on the singing of guardian spirit songs over the many days of the dance. Singing begins in the evening of each day and goes on until dawn. The ceremony also includes the activities that go on during the day, such as feasting, sweat-lodge ceremonies (to heal, cleanse, and pray), giveaways, stick-game gambling, and storytelling. At the heart of the Winter Dance ritual is the relationship of the individual to the guardian spirit. This exchange between humans and spirits creates and reenacts the time of the old stories of myth, or **cosmogony**, in which the universe was created. The Salish moral imagination is set forth in the symbols of this creation story that is believed to bring new life to the group and to cause a rebirth of plants and animals. Thus, the relationships between humans and guardian spirits form the core of the Salish synthetic ethics in which stories, songs, and symbolic actions bind individual, group, and bioregion together to create a holy togetherness and a **spiritual** understanding.

Celebrating universal interdependence. As a world rebirth ceremony, the Winter Dance calls the spirit powers of the bioregion into a give-and-take relationship with the human communities. This ceremony makes clear how minerals, plants, animals, and humans depend upon one another; it does this through the songs that are sung by those who have had visions of these spirits in special places of the bioregion. There is no actual telling of the creation myths during the Winter Dance. But during the days and before the evening and all night ceremonies, individuals are urged to tell stories. Coyote stories are especially popular at these times. While there is no single explanation of the creation of the world among the Salish people, the many Coyote stories have parts that describe the events that formed the world in the time of mythic beginnings.

The center pole, a symbol of the bioregion, is set up in the middle of the dance hall. Through songs and giveaways the people show their moral imagination. The singers are believed to go through a spirit sickness because of their closeness to the creative powers. The singers go to that center pole to sing, speak in moral appeals to the group that has gathered, and give gifts just as the ancient spirits of myth gave to humans. While dancing around the pole to the songs of those who see visions, the dancers are said to be like the animals who "are moving around" during the snows of the Winter Dance season. The very idea of the Winter Dance as animals moving about the land is seen as having moral force in Salish thought. More than just single ceremonies or symbolic acts, the Winter Dance is understood as bringing a person and a community into the moral order set up during the time of creation, when the plants, minerals, and animals decided to give their bodies to humans for food.

The Elk medicine dance (Glacier National Park, Montana).

shaman a religious specialist, similar to a priest, who uses spiritual practices and special ceremonies to cure the sick, reveal sacred mysteries, and affect events

Health, Sickness, and Healing

Knowledge of health, birth, and death among certain native North American peoples came about as a result of their exploration of their bioregion, and from historical contacts with other native peoples long before the arrival of Europeans. One very old religious practice, that of the healer, or **shaman**, still keeps up the traditional knowledge of bioregions collected over centuries of historical change. Native American peoples probably brought healing practices with them when they moved across the continents from Siberia as long as 40,000 years ago. The shaman used rituals for mental and spiritual healing. Others who treated illness used herbs.

Five hundred years of contact with Europe and Asia brought native peoples "virgin soil epidemics," diseases against which they had no natural defense. These diseases caused widespread destruction of Native American populations. This at first challenged and then decreased the ceremonial power of Native American shamans during the seventeenth, eighteenth, and nineteenth centuries, but it did not lead to the disappearance of these ceremonies. As outbreaks decreased, traditional practices were often given full credit for curing them. At the end of the twentieth century, traditional healers were often found working together with medical doctors in many places where Native Americans live.

Medicine men. Ritual specialists who can identify a disease, treat illnesses, and guide the dead are found in all traditional native North American settings. In many farming communities, these specialists formed priesthoods that handed down traditional stories and ceremonies that healed specific illnesses. Often, specialists in dreams, visions, and spiritual travel to other-than-human places were believed to have obtained knowledge of the human body that could not be learned by simply observing things.

Native survivors of the gathering-and-hunting groups of North America usually approved of individual shamans. Unlike priests,

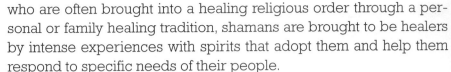

who are often brought into a healing religious order through a personal or family healing tradition, shamans are brought to be healers by intense experiences with spirits that adopt them and help them respond to specific needs of their people.

The meaning of disease. Disease in a traditional Native American group is usually thought to come from breaking a spiritual law, doing something forbidden, having an object "shot" into a diseased person by witchcraft, or the loss of life-force or soul. A list of forbidden acts often makes up the major part of an ethical system involving hundreds of rules for the treatment of living things, handling the remains of dead creatures, and ways of living with the spiritual powers in the bioregion. The Koyukon people of Alaska have a complex system of rules and laws called *hutlanee*. Disease and death can result from breaking these rules and destroying the natural balance of *sinh taala*, the power of the earth. Koyukon shamans, *diyinyoo*, know the spiritual powers that live in the bioregion and use their power to identify disease, to treat illness, and to bring back *sinh taala*, the most important part of their medicine. Shamans and elders teach the wisdom needed to bring back the power of the earth and to meet death with knowledge of the paths to those places in the bioregion where the dead one will live.

Disease that comes from objects entering the body, or witchcraft, often is based on a worldview in which balance or harmony between one's body and the local bioregion has been broken on purpose by a hateful person. Among the Dine/Navajo, the health of a person is not a separate case but a matter of the whole world of living things. The "beauty," *hozho*, built into the world can be put out of order by evil acts of witchcraft. Creation ceremonies of great beauty, called chantways, are done by ritual specialists, called singers, to bring back peace and order to the sick person's body by getting rid of the object that has entered or by getting back lost life-force.

Healing ceremonies. In the Dine/Navajo creation story, the basic source for the chantway stories, the beauty of the earth is called upon in the following chant to restore health:

> Then go on as one who has long life
> Go on as one who is happy
> Go with blessing before you
> Go with blessing behind you
> Go with blessing below you
> Go with blessing above you
> Go with blessing around you
> Go with blessing in your speech
> Go with happiness and long life
> Go with mystery.

▲ A Navajo medicine man creates a sacred sand painting.

Through these repeating words, the singers raise holy power and control the inner forms of themselves, of the sick person, and of

▶ A Navajo medicine man chants before a sand painting while healing a sick child.

the spiritual powers in the bioregion that have been called upon in the sandpainting ceremony. The singer brings back the well-being of the one sung over by bringing the sick person into the healing setting.

Current Ethical Viewpoints

Major ethical issues involving native North American peoples center on the following three areas: bones of ancestors, religious freedom, and self-governance.

Ancestors. The passage of the Grave Protection and Repatriation Act of 1990 helped to slow the destruction of old Native American gravesites. The listing of Native American objects in major museums also has made it more likely that sensitive religious material will be returned to the native peoples from whom it was often improperly taken.

Religious freedom. Serious questions of trust between the American government and Native American peoples have come up in a number of court cases in which native peoples' religious freedom has been limited. The history of how the American majority's culture and laws has been against Native American lifeways was stopped for a while with the passage in 1978 of the American Indian Religious Freedom Act. But since then a number of Supreme Court cases have challenged the freedom of Native American lifeways and showed an unwillingness to honor their holy relationships to land.

Self-governance. The worldwide voices of native peoples were raised in the latter part of the twentieth century, warning of the dangers to the environment and protecting the treatment of native

Traditional dancing at an inter-tribal pow-wow.

peoples in parts of the world that have become newly desirable. In the United States and Canada, native peoples, after having been pushed onto reservations and areas away from the majority populations and culture, now control resources and undeveloped land. Native American peoples have increased their close contact with other native peoples around the world in an effort to protect themselves from projects harmful to their bioregion. Such projects include dams that produce energy and toxic waste dumps that destroy the environment. Meetings such as the United Nations Earth Summit in Rio de Janeiro in 1992 and the meeting called "Changing Ecological Values in the Twenty-first Century" in Kyoto, Japan, in 1993 have included native North Americans. Meetings have also brought together people from the world's religions to talk with elders from Native American lifeways about their traditional environmental ethics.

The rebirth of the native nations. The remarkable revival of native North American peoples in the late twentieth century, after 500 years of being kept down, resulted from a complex process. The knowledge passed down in traditional ethics played a major part in their survival. Often looked down upon as superstitious or called primitive, the insights of native peoples are now seen as ways of knowing based on a close relationship with local bioregions. This is a far cry from a time when the "nobility" of Enlightenment reason was compared with the "ignoble" state of native thought. For traditional native North American peoples, the world is alive and, far from being a random group of things, is seen by some as their Mother and by many as a community of knowing subjects. Not just a branch of knowledge, Native American thought in bioethics brings one to the heart of a way through life.

Related Literature

Leslie Marmon Silko, in the novel *Ceremony (1977)*, describes a Native American, Tayo, who is a member of the Laguna tribe. He has returned from fighting in Vietnam suffering from "battle fatigue." At the Veterans' Administration hospital, white doctors tell him to avoid "Indian medicine" and to distrust the traditional native American methods of healing. His illness gets worse. He suffers chronic nausea, hallucinations, and depression. Finally his grandmother insists that Tayo see a Laguna healer, who helps Tayo come to terms with himself and his heritage. The white man's medicine, with its implicit racism and colonial attitudes, nearly destroyed Tayo, but he starts to heal as he hears the stories of his own people and reclaims his own tradition.

PROTESTANTISM

Protestant Christianity began during the sixteenth century as a reform movement within the Roman Catholic Church. The reformers protested against certain things about the Church, but the proposed changes were rejected. As a result, Protestant (from "protest") churches were formed, first in northern Europe and then in North America. During the last 400 years, Protestant churches have been founded all over the world.

The Origin of Protestantism and Its Beliefs

Protestantism was a protest against the way that the **medieval** Christian Church was run by its leaders. The word Protestant comes from the Latin word, *protestari*, which means "to give testimony." Protestantism began with protest, but it is also a movement of religious belief and **testimony**.

Martin Luther was a young German monk of the Roman Catholic Church, who in 1517 led the protest against the medieval Church, whose authority was the pope in Rome. At that time, people who had sinned could pay money to the church for **indulgences** and grants of forgiveness of their **sins** from the pope. Martin Luther believed that God's forgiveness could not be bought with money paid to the Church. He had hoped to discuss this problem with Pope Leo X. Instead he found himself in an argument with the pope over basic questions about the beliefs of the Church. In 1520, Luther wrote three articles against the practices of the Church under the pope's leadership. Without intending to form a new church, Martin Luther's ideas founded Protestant Christianity.

God's sovereignty and Christians' freedom. Luther wrote about two important ideas. One was God's sovereignty, or absolute rulership over all, and the other was every Christian's right to freedom under God. His articles also said that God was a **gracious** God. When Luther said that God has sovereignty, he meant that God was all powerful and chose to be righteous and gracious. God's decision was purposeful and freely made. Luther wrote that God decided to be known to the world through Jesus Christ and through writings about the life of Jesus Christ that are called the New Testament Scriptures. God's free choice to be these things gave Christians freedom as well. Sinful human beings could freely choose to have Christ as their lord, and to live in the righteousness and **grace** of God. They could also choose to serve God and to serve their neighbors. Luther's proposals about God shed light on Christianity. He wrote that a Christian is a servant to everyone.

Faith and works. In his articles, Luther also objected to the Church system of **sacraments**. He thought it was wrong that popes and priests could give or withhold the grace, or forgiveness, of God. Luther believed that God's grace is free and available to everyone. The system the Church used made God less free and made Christians less free as well.

---

medieval relating to the Middle Ages (500–1500 C.E.)

testimony a public declaration of religious experience

indulgence a pardon for sins committed and a removal of the punishment for them

sins acts that are religiously or morally wrong and often bring punishment or require atonement and making amends to those wronged

gracious compassionate or merciful

grace God's assistance to humans to help them shape their lives, which God gives them out of his compassion and mercy

sacrament one of several Christian rites (such as baptism) believed to have been established by Jesus as a means or symbol of divine favor

penance an act that shows sorrow or repentance for sin; a sacrament in some churches that includes a private confession, absolution, and an act of repentance assigned by the confessor

canon a rule or accepted belief decreed by a church council

excommunicate to exclude from a religious community as a way of punishing a transgression—usually heresy

baptized having undergone baptism, a Christian sacrament; it is a ceremony marked by a ritual use of water and admits the recipient into the Christian community

ministry the body of ministers of clergy of a religion

heresy an opinion or doctrine contrary to accepted church beliefs; the sin of holding such a doctrine

An example of a sacrament was **penance**. Luther thought that the way penance was given by priests and received by the person confessing a sin was wrong. The priest listened to a person's confession, named the kinds of sin, and decided what kind of punishment or penance was necessary for forgiveness. **Canon** law allowed priests to make judgments in this way. Priests used confessional manuals to help them recognize and deal with different kinds of sins. Because of this practice, the Church had very strong control over every part of life, including physicians and the care of sick people.

For Luther the main sin of the Church was that having priests hear and forgive sins seemed to show that the Church did not trust in God's forgiveness. Luther wanted to warn the Church against accepting money for giving God's grace to sinners and against making judgments about other people's sins. He warned the Church that it was serving itself instead of serving others.

The "priesthood of all believers." Luther also believed the organization of the Church should change. He did not think it was right for a single person, a pope, to have so much power. Luther objected to the pope, because he believed that Jesus Christ was the true leader of the Church. He felt that the Church system, which put the pope at the top, took freedom away from Christians. It also made Christ's power over the Church less important. In comparison with Christ's authority, the levels of power within the Church were really all the same. Pope Leo X **excommunicated** Luther for his statements against the Church.

According to Luther, anyone who was **baptized** had "put on Christ," and all Christians could represent Christ to others. He felt that all baptized Christians were like priests. This meant that each Christian could act in a Christian manner toward others and call on God for help if someone needed it. Luther called this the "priesthood of all believers." It gave respect and importance to all Christians and also made it clear that everyone was part of a Christian community. Only Christ had ruling power in the church. Everyone else was part of Christ's **ministry**.

The diet at Worms. The German Emperor Charles V called Luther to appear before a group of leaders, the diet, assembled at a city called Worms in 1521, because of his beliefs. The Emperor ordered Luther to deny everything he had been saying, but Luther refused. He said that he was following the law and word of Scripture. The only thing that would make him take back his words was something in the Scriptures themselves or clear reason. He could not with a good conscience do what the Emperor asked. "Here I stand," he said, "I cannot do otherwise."

Luther's preachings were seen as **heresy**. In 1529, at the diet of Speyer, Charles V tried again to stop Luther from preaching his beliefs. The diet of Speyer was made up of 400 estates in Germany, and leaders from 19 of them took Luther's side. These leaders said

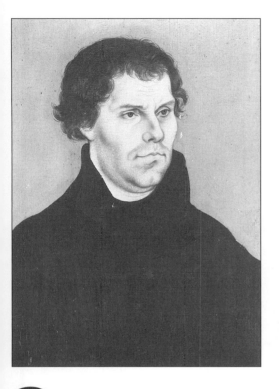

Martin Luther

that they too believed in the authority of Scripture and in God's supreme power. They also believed that God had power over the pope and the emperor. These nineteen representatives and all the other people who believed Luther's preachings came to be called Protestants.

Many Protestants

Protestants believe that the only legitimate source of spiritual truth is the Bible. However, different interpretations of the Old and New Testaments and different opinions about church organization, infant and adult baptism, and the meaning of Communion, among others, led to the formation of many different churches in Protestantism. In the United States these include Baptists, Congregationalists, Episcopalians, Lutherans, Mennonites, Methodists, Pentecostalists, Presbyterians, Unitarians, and a large number of smaller groups. As may be expected, there is little agreement among Protestants on issues of medical ethics.

Although Luther was only a young monk, he was joined in his beliefs by scholars and religious thinkers from all over the world. Some of the people who agreed with Luther included Phillip Melanchthon in Wittenberg, Ulrich Zwingli in Zurich, John Calvin in Geneva, Menno Simonsz in the Netherlands, and Thomas Cranmer in England. But not everyone agreed completely about the kinds of changes that were necessary. Although together they had rebelled against the Church and the pope's authority, the differences between them eventually caused them to split from each other. The people who followed John Calvin broke away and became the Reformed (or Presbyterian) Church. Others formed groups that were more extreme in their disagreement with the Church, for example, the Mennonites. One group whose beliefs were less extreme than Martin Luther's formed the Anglican Church in England.

Protestant Views on Medical Ethics

In 1960, Joseph Fletcher published a book called *Morals and Medicine.* It was not the first book to present Protestant ideas about medical ethics, but it was an important book. Fletcher's book was set against the tradition of Roman Catholic **casuistry**. But Fletcher used only certain parts of the Protestant religion in his book. One aspect of Protestantism he looked at was human freedom. He also discussed the belief that knowledge means mastering and controlling nature to help other human beings. Fletcher also presented Protestantism as having the independence of a conscience that is influenced by Scripture. In Fletcher's book, this very important belief was reduced to the commandment that says we must love our neighbor.

casuistry the interpretation of religious principles or doctrines to make decisions about specific instances of conscience, duty, or conduct

During the next ten years, Protestants were concerned about other kinds of issues, such as civil rights and the problems of war. Then in 1970 Paul Ramsey wrote several essays about genetic control and abortion. Protestants began to think again about medical ethics. Ramsey also wrote a book called *The Patient as Person: Explorations in Medical Ethics*. After this book was published, many other people began once more to think and write about these issues. Over the next ten years, many Protestants wrote about medical ethics.

It is hard to say any one thing about the books that were written, because the authors had different ideas and opinions. They were influenced by many different traditions over a long period of time. These traditions include the beliefs of the sixteenth-century followers of Martin Luther, those of the Puritans in the seventeenth century who adopted the ideas of the philosopher Francis Bacon, and ideas from the eighteenth-century **Enlightenment** and "awakening." Some of these authors think medicine is wonderful, while some do not trust it at all. Despite all these different influences and beliefs, there are some principles all these Protestant writers agree on.

God's sovereignty. Protestants believe that God is all-powerful and is the only real authority and ruler. They believe in God's sovereignty over popes and kings, and over the laws developed by churches and those developed by governments. For Protestants, God alone is God. Only God deserves human trust and confidence, and only God deserves human faith and belief. This system of belief puts God at the center and nothing can be compared to God. Protestants feel that the greatest danger to their belief system happens when someone or something is treated in the same way that God is treated. In Protestantism, **idolatry** presents a great mortal danger.

Enlightenment a movement in eighteenth-century Europe that used reason as its guiding rule

idolatry the worshiping of something as of God

▶ Martin Luther (1483–1546) is surrounded by John Calvin, Philip Melanchthon, and other sixteenth-century church reformers.

"Look to your health; and if you have it, praise God, and value it next to a good conscience; for health is the second blessing that we mortals are capable of."

—Izaak Walton (1593–1683), English writer

Protestant churches are deeply divided over the issue of **abortion**. Liberal Protestants have strongly defended a woman's right to an elective abortion, whereas more conservative Protestants have led the pro-life movement in opposition to abortion.

abortion the deliberate or spontaneous early ending of a pregnancy, resulting in the death of a fetus

fetus an unborn baby

Concern about this danger can be seen in Protestant beliefs about medical ethics. For example, Protestants believe that giving life the same respect that is given to God is a form of idolatry. The theologian Karl Barth wrote that life deserves respect because it is a gift of God, but he also said that the respect given to God must always be greater than the respect given to life. Protestants believe in the absolute authority of God. This means that there might be a situation where respecting God means taking a life. It also means that there might be times when prolonging a life would not be respectful of God's will.

For Protestants, technology creates a strong threat to the sovereignty of God. Many Protestants have expressed the feeling that our society has too much confidence in technology. Then there are Protestants who think it is wrong to worship "Mother Nature" over technology. They believe that we must always take advantage of technology if it can help human beings.

The patient's autonomy. Human freedom was as important to Martin Luther as the rulership of God. Many of the Protestants writing about medicine discuss the importance of freedom. Protestants have contributed the most to medical ethics by discussing the roles and rights of all human beings. A big part of Protestant tradition is the belief that all people should be free to make choices and decisions. This is why people should not be ruled by those in positions of power, whether in the church or in the medical field.

The focus on freedom in the works of many Protestants who have written about medical ethics can be seen in the writings of Joseph Fletcher, who emphasizes the importance of having "respect for persons" and the idea that we should never use others for our own purposes. Fletcher feels strongly about people's rights, and he also believes in the importance of informed consent. That means that we should have all the information we need before we agree to something, and that we agree to it because we want to. Ramsey's themes in his book *The Patient as Person* are similar to Fletcher's. Ramsey is concerned that medicine and medical research do not always treat patients as persons. He feels that doctors are sometimes too concerned about benefits to research. Like Fletcher, he believes in "free and informed consent," because it helps to make sure that the patient is treated with respect and creates a caring and loyal relationship between patient and doctor.

But Fletcher and Ramsey do not agree about everything. They have different ideas about who can be considered a person and about whether having respect for a person's choices is enough. According to Fletcher, a "person" is someone who can think, learn, and make choices. Ramsey has problems with this definition. For Ramsey, anyone with a physical body is a person, and he believes that Fletcher's idea excludes impaired individuals. Fletcher's definition would say that **fetuses**, newborn babies, or a person in a coma

sacred dedicated or set apart for the worship of God or a god; holy

who can no longer function normally are not persons. Since they are not persons, according to Fletcher, they can be treated in a way that is most helpful to others. They do not have to be respected. Ramsey thinks this is wrong. He believes that fetuses and newborns are human creatures and are **sacred**.

Calling and profession. In the sixteenth century, when Protestantism was a very young religion, the idea of a "calling" or "vocation" allowed doctors to escape the control of the church. The idea that a doctor was supposed to do that kind of work gave medicine a new kind of respect. But later, when doctors had become independent, Protestants continued to regard this work as a calling. In general, Protestants want doctors to see their jobs not as technical professions but as a way of responding to God's grace. Protestants want doctors to feel they are doing a service that has God's approval.

▶ Friends lay hands on a teenager during a prayer at a charismatic church service.

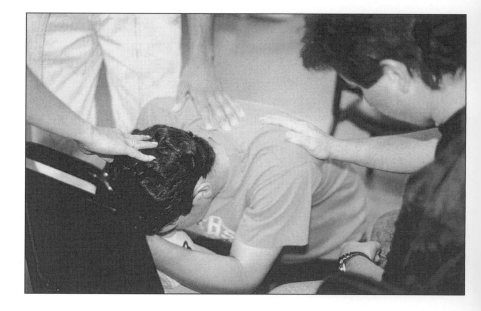

One Protestant writer, Walter Rauschenbusch, hoped that doctors and nurses would feel that by healing the sick, they were working for Christ. Rauschenbusch was a minister and social reformer in New York City in the early twentieth century. Protestants like Rauschenbusch felt that doctors must not work just for money. He felt they should have a sense of "divine mission" and should show their faith in God by helping the poor.

Some Protestants say that medical work can be a "calling," because they believe that helping and healing the sick can serve the cause of God. Protestants also make this point when they write

covenant a moral obligation of special loyalty

about the idea of the **covenant**, a kind of promise or agreement. This idea was important to Ramsey. He felt that informed consent was a sign of a covenant made between doctor or researcher and patient. But Ramsey did not explain the religious philosophy of covenant and instead focused on covenants between persons.

Protestants have added variety to ideas about right and wrong in medicine and medical practice. But they have also contributed their ideas about the importance of God's rulership over everything and the freedom and rights of patients. Protestantism also shows that it is important to accept that sometimes questions of right and wrong cannot be answered. Finally, Protestantism contributes a powerful sense of vocation and shows us that scriptures and religious writings can support, guide, and help people who are suffering.

ROMAN CATHOLICISM

The Catholic Church firmly condemns **euthanasia**, although it is very accepting of treatment refusal and withdrawal when patients are terminally ill. Euthanasia, however, goes beyond allowing nature to take its course. Euthanasia involves active killing. The Catholic Church also condemns assisted suicide, and has opposed legalization of this practice in the United States and in the world.

euthanasia mercy killing

sin an offense against religious or moral law

spiritual relating to the nonphysical intelligence or feeling part of a person; having to do with sacred matters or religious values

sacrament one of several Christian rites (such as baptism) believed to have been established by Jesus as a means or symbol of divine favor

anointing the ceremonial application of oil to a person's head to show that he or she has been chosen by God to do a task

Since its beginnings, Roman Catholicism has been concerned about people's bodies as well as their souls. Catholic thought includes ideas and teachings about what is right or wrong regarding all things that affect the human body, including how people should be treated when they are ill. Beginning in the 1960s, people from all parts of society became concerned about the ethics of some scientific research and medical procedures. Roman Catholicism had a history of dealing with these subjects. Medical ethics and a concern about the body have been part of Church teachings for a long time.

The Duty of Care

The Roman Catholic tradition has its beginnings in the very old writings of Hebrew and early Christian Scriptures. One important idea in those writings is that the sick should be taken care of. The Christian point of view about sickness is that God creates and gives us life, and He is the one who heals us. Sickness and death are due to the power of **sin** in the world. Sickness is also believed to come from human frailty. Roman Catholics believe that we must try to heal ourselves and get help if we are sick, but in the long run it is important to understand that everyone will die. Roman Catholics believe that when we are sick and we feel pain, we are overcoming evil. This kind of suffering also lets us know how Jesus Christ felt. When we are sick and in pain, we can begin to share in the suffering, death, and resurrection of Jesus.

The Roman Catholic Church believes that there are different kinds of sickness, and it also believes in different kinds of care. One kind of sickness is **spiritual**. The Church takes care of spiritual sickness through the **sacrament** of **anointing**. This practice is one of seven Roman Catholic sacraments. For a long time, this sacrament was given only to a person who was dying. In that case, the sacrament is called extreme unction. But recently the Church decided that this sacrament should be given to the sick as well as to the dying, which was the original idea. This sacrament is related to Jesus, because it celebrates his power to heal the sick. It also celebrates God's power to change sin, sickness, suffering, and death

resurrection the rising of Jesus from death; or, the rising of all the dead before the final judgement

mediation intervention between two parties to help their communication or to achieve a certain result

Ethics and Law: Care and Compassion

repenting feeling regret for a sin and choosing to amend one's life

penance an act that shows sorrow or repentance for sin; a sacrament in some churches that includes a private confession, absolution, and an act of repentance assigned by the confessor

penitence the state of feeling sorrow for sins or faults

absolution a removing of the burden of having committed a sin

adultery sexual relations that involve a married person and someone to whom he or she is not married

abortion the deliberate or spontaneous early ending of a pregnancy, resulting in the death of a fetus

itself through Jesus Christ's death and **resurrection**. Prayers for healing are a part of the sacrament of anointing and part of Christian life in general. But those who put their faith in this religion also know that not everyone will be healed.

In general, Christians believe that God often answers prayers through people, and Catholics believe in this idea strongly. This principle is called **mediation**, and it has a great influence on Catholic beliefs. For example, when it comes to healing the sick, the Church encourages the help and care of doctors and other healers. Although everyone will die in the long run, Catholics feel it is good to serve God by helping sick and dying people. Catholics believe that part of human life is taking good care of oneself and others. Sometimes there have been conflicts between the medical field and the Roman Catholic Church. But overall, the Church has been encouraging to doctors and supports their work.

The Catholic Church has played a big part in the creation and staffing of hospitals. Many Catholic men and women have made it their life's mission to help those who are sick and suffering. In addition, the Church praises and shows a great deal of respect for doctors, nurses, and others in the field of health care.

Morality and Medicine

The field of medical ethics has grown for a variety of reasons. Being a Catholic means doing good deeds and helping others. The Church tells people that they should examine their actions to decide if these were sinful or good. Believing in God is not enough to make you a good Christian. Doing good things for others is important as well. In the Roman Catholic religion, the practice of **repenting** for your sins is important. People who have sinned in any way must pay for their sins through acts of **penance**. Because this **penitence** is very important, Catholics need to know which kinds of behavior are sinful and which are not.

Penitential books. From the seventh to the twelfth century, the *libri poenitentiales*, penitential books, were written. They were written to help priests understand the sacrament of penance, and they describe the process that needs to be followed. First, a person confesses his or her sins to a priest, who decides how serious the sin is and what kind of penance needs to be done. Then the priest gives the person **absolution** in God's name. In these books, sins were arranged by subject and the correct penance that a priest should give for each sin was explained. The list of sins included stealing, lying, cheating, and **adultery**, but also birth control and **abortion** and other matters related to sexual behavior and marriage.

Living a moral life. During the thirteenth century, ideas about the Catholic religion developed even more. Thomas Aquinas, who died in 1274, was considered the most important religious thinker at that time. His ideas were accepted by many people. His

natural law a principle or set of laws that derive from nature, right reason, or religion and are ethically binding

Fertility and Reproduction: Abortion, Fertility Control

Thomas Aquinas (1225–1274), Italian religious and philosopher, was canonized in 1323. He is shown on the Demidoff altarpiece as a man of both church and science.

approach to Catholicism is very systematic and orderly. In the second part of his famous work, *Summa theologiae*, he writes about how Christians can live a moral life. Using his systematic approach, a moral Christian life has three parts: the human being related to God as ultimate end, the human being as the image of God insofar as he or she is capable of self-determination, and the humanity of Christ as the human way to God. Thomas also wrote about the theory of **natural law**, and he used logic and reason to understand human behavior and develop ideas about moral right and wrong. In general, the Catholic Church takes the approach to morality that Thomas Aquinas took.

A treatise on medical ethics. A few hundred years later, in 1621, Pablo Zacchia, a doctor who lived in Rome, wrote a multivolume book called *Questiones medico-legales*, "Medical Legal Questions." He was the first Roman Catholic thinker to write about medical ethics. His book covers many subjects including age, birth, pregnancy, death, mental illness, poison, the inability to have children, plagues, contagious diseases, virginity, rape, fasting, cutting off or injuring parts of the body, and the sexual relationship between husband and wife. Zacchia tried to connect religion with medicine and the law, and his books have had a great deal of influence. They teach the priest and pastor medical information they need, and help doctors think about the moral and legal issues they face.

Pastoral medicine. During the twentieth century, Catholic writers continued to be interested in the relationship between medicine and religion. All over Europe, writers and thinkers were publishing books about pastoral medicine and medical responsibilities. The books on pastoral medicine tried to understand and define the healing jobs of priests and pastors. There were also writings about doctors' moral responsibilities and the rights and wrongs of medical research.

In 1949, a German writer named Albert Niedermeyer wrote an important work on pastoral medicine. This book took a fresh look at topics that had been discussed in earlier publications but also dealt with questions that arose in modern times. For example, Niedermeyer discusses psychiatry and psychotherapy. It was during this period that the field of medical ethics focused on the moral questions that doctors, nurses, and staffs of Catholic hospitals had to think about. Finally, the Catholic Hospital Association in the United States wrote rules called "Ethical and Religious Directives for Catholic Hospitals." It was first written in 1949, and rewritten in 1955.

Natural Law and Free Will

In the United States especially, people began to understand that topics Niedermeyer wrote about belonged to the field of medical ethics. For Roman Catholic thinkers, medical ethics had two main themes. The first was natural-law methodology and the second was the

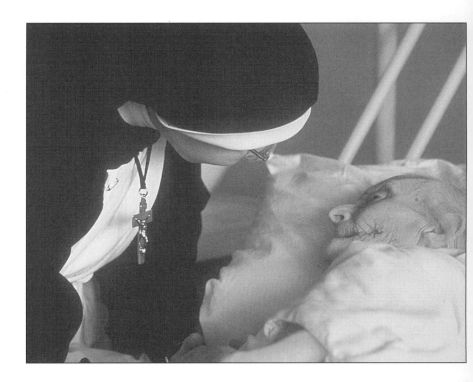

▶ A nun comforts an elderly patient in a hospice.

The Catholic Church in the United States published its first Ethical Directives for Catholic Hospitals in 1949. The text was revised in 1954, 1971, 1975, and 1994. Following are some excerpts from the 1975 guidelines.

Directive 10 states, "The directly intended termination of any patient's life, even at his own request, is always morally wrong." Directive 11 affirms, "From the moment of conception, life must be guarded with greatest care. Any deliberate procedure, the *purpose* of which is to deprive a fetus or an embryo of its life, is immoral."

importance and authority of Church teachings. Before the twentieth century, the Catholic religion taught that moral wisdom and Christian knowledge came from two sources: one was reason and natural law, and the other was belief in God and in the Scriptures. But the textbooks on medical ethics that were written during the twentieth century mostly focused on natural law and not on biblical writings.

Eternal law. Catholics also believe in the eternal law, which is based in the belief that God has made the world as it is and should be and that nature should be respected. The eternal law comes out of God's being and existence. For Catholics, natural law is the way the eternal law works for all of God's creations. Thinking creatures do not simply follow God's directions, they also use their minds to reason and take an active part in following their path. Catholics believe that those with right reason can see that human beings have three natural ways of being: one is having the desire to continue living, which is something we share with all living things; another is having the need to have children and teach our children, a need humans share with all other living creatures; and finally, we have the desire to live with other thinking creatures in a community. Through these three elements, natural law helps human beings use their minds and their ability to reason to follow their natural path in life.

Natural law. Natural law states that everyone knows instinctively that people should do good, stay away from evil, and do things according to reason. Because we are thinking creatures, we can also understand more complicated versions of these basic ideas. For example, we know that staying away from evil means that we should not steal or lie or cheat.

On April 11, 1973, immediately following the U.S. Supreme Court decision legalizing abortion (*Roe* v. *Wade*), the National Conference of Catholic Bishops declared that the "opinion of the Court is wrong and entirely contrary to the fundamental principles of morality. . . . Wherever a conflict arises between the law of God and any human law we are held to follow God's law." The Bishops then declared that no Catholic hospitals or other medical institutions could perform abortions, and that Catholics working as health-care professionals in non-Catholic setting should refuse to participate in abortions.

uterus the organ in women in which the fetus develops; also called the womb

encyclical a letter from the pope to all the bishops of the church or to those in one country

Authority and Power of Church Teachings

The second major part of Catholic medical ethics deals with the authority and power of Church teachings. According to the beliefs of the Roman Catholic Church, God gives the pope and the bishops and all the councils of the Church the power to teach those who follow the Catholic religion. God wants them to act in his name and teach Catholic believers about faith in God and about right and wrong.

During the nineteenth and twentieth centuries, many people were writing and thinking about medical ethics, and at the same time, people were paying more attention to the teachings of the pope. For example, between 1884 and 1902, many people wrote to the pope with questions about abortion. The pope then made it clear that abortion was wrong in some situations that had not been written about before. The Holy Office wrote that even if a pregnant mother's life was in danger, the fetus (unborn baby) could not be killed. The Office also decided that if the fetus was ectopic (not inside the mother's **uterus**), it still could not be removed from the mother's body. **Encyclicals** about marriage written by Pope Pius XI (*Casti connubii*, 1930) and by Pope Paul VI (*Humanae vitae*, 1968) also clearly stated that it was wrong for married couples to use birth control.

While Pope Pius XII was in the Holy Office from 1939 to 1958, he spoke about questions of medical ethics a number of times. His declarations made it clear that the authority of the Church's teachings is a very important part of the Catholic religion. Pius XII showed a great deal of interest in the subject of medical ethics. He often spoke to groups of medical doctors and researchers. Some of the subjects he talked about included the duties of people in the medical field, giving blood, artificial insemination, birth control, sterilization, abortion, medical experiments, genetics, painless childbirth, transplants, death, and the different ways science had found to keep people alive longer. The pope's interest in these questions caused many other thinkers and writers to take an interest as well. This is one reason the field of medical ethics developed so quickly during this time.

Moral Principles in Catholicism

The Catholic tradition determines natural law by first explaining and establishing its principles, and then looking at specific cases with these principles in mind. Through these specific cases, Roman Catholic beliefs about medical ethics developed several important ideas.

The right to life. One important principle in Roman Catholicism is the right to life. Catholics believe that life is a gift of God and anyone or anything that is given life has the right to keep it. This right cannot be denied by any government or by another human being. The right to life cannot be renounced or given away by anyone, not even by people who try to kill themselves. But Roman Catholics understand that sometimes it is necessary to kill someone else in order to defend oneself. The Church also understands that during war one person may have to kill another. It also recognizes **capital punishment** and accidental killing. The basic idea behind the right to life says that it is never right for a person to decide to kill an innocent person.

capital punishment punishment by death

Right of use, or stewardship. Another part of natural law is the belief that we do not own our bodies. We only use them while we are on Earth. As thinking creatures we have the right to use our bodies in ways that God would approve of, in ways that are natural. But this belief means that the Catholic Church does not feel that people have the right to hurt or destroy their bodies. This is called the principle of stewardship. According to this principle, people must take care of their physical selves. It is this part of natural law that makes it okay for people to have surgery if they need it in order to be healthy.

Sexuality and procreation. Sexual relations have two purposes. One purpose is to express the love between husband and wife, and the other is to create children. This is why the Church believes that sterilization and birth control are wrong. It is wrong for a person to try to change the result of the sexual act. It is an act that God created for a purpose, and it is not up to individuals to change it.

Fertility and Reproduction: Fertility Control

Specific Questions

The beginning of life and birth. Catholics are not absolutely certain about when life begins, but for the most part they believe it begins with **conception**. The Church feels that the fetus inside the mother's body is helpless and must be protected, and Catholic religious beliefs try to protect all fetuses. The Church also states that

conception the act of becoming pregnant; the moment when the egg is fertilized by the sperm

A nun carries the elements of Communion to women in a maternity clinic in Cité Soleil, a slum of Port au Prince, Haiti.

direct abortion is always wrong. If there are very good reasons, indirect abortion might be permitted. One common example of indirect abortion is when a woman's uterus is removed because she has cancer of the uterus.

Death and dying. Because Catholics believe that we do not own our bodies and are only in charge of caring for our body during our life, they also object to active euthanasia. But Catholics feel that it is good to give dying people painkillers to ease their suffering—even if the painkillers cause death to come more quickly. Catholics also believe that there is no need to go to extremes to save a patient's life when treatment is burdensome and death inevitable. This is clearly different from active euthanasia. Catholics are always supposed to care for the health of other people, even when the situation is unbearable or hopeless. But it is not good to keep a patient alive if it means that the medicines, treatments, and operations that would be needed are very expensive, painful, or inconvenient. From this point of view, it is right to choose not to use a respirator to keep a person alive for a few hours or days. It is also right to shut off a respirator that is already in use.

Other questions. In 1968, Pope Paul VI wrote the encyclical *Humanae vitae*. In it he repeated the Church's objection to birth control; and for the first time in the history of the Church, some Catholic thinkers publicly disagreed with the pope. They said that all Catholics have the right to disagree, because this particular teaching could be seen as imperfect. They argued that some questions of right and wrong rely more on human reasoning and thinking than on faith in God. Catholics believe in the authority of Church teachings, but that belief can be overturned if there is evidence that the authority is wrong or if there are other good reasons. Since then, other Roman Catholic thinkers have publicly disagreed with other Catholic ideas about medical ethics. The Church says it can tolerate disagreement of this kind; but there have been situations where the Church has punished certain religious thinkers for their disagreement. Some Roman Catholics believe that such disagreement goes against one of the most important parts of the religion—the authority and power of Church teachings.

While the Catholic Church views **homosexuality** as immoral, it nevertheless insists on compassion for all those suffering from AIDS. It is important, according to Catholic teachings, that we care about people with AIDS because all life is sacred and merits compassion.

homosexuality sexual attraction or activity between persons of the same sex

A strong criticism of Catholicism comes from population control advocates who blame the Church for overpopulation in the Catholic countries of South and Central America, as well as in other areas of the developing world. The Catholic Church has responded with the claim that there is no overpopulation; the real problem is that wealthy nations do not come to the assistance of poor ones.

Related Literature

Graham Greene (1904–1991) was an English writer who converted to Roman Catholicism when he was twenty-one. In *The Power and the Glory* (1940), an alcoholic priest risks his own life to **anoint** a dying gangster, showing how even sinful people can be saints.

anoint apply oil to a person's head as a sign of blessing or sacred designation

SIKHISM

guru a religious teacher and spiritual guide

meditation spiritual introspection or the contemplation of a religious truth, often practiced in order to reach a higher state of understanding or realization

mantra word(s) or sound(s) used as a form of invocation or incantation

transmigration the passing (of a soul) from one state of existence to another; also, to pass at death from one body or being to another

Sikhism began with **Guru** Nanak, who lived from 1469 to 1539. Guru Nanak was born a Hindu in the Punjab, a region in northern India that is still home for the vast majority of Sikhs. The word *sikh* means "learner" or "disciple." The Sikh community numbers approximately 16 million people.

Origins and Teachings

According to Sikh history, Nanak was the first of ten personal gurus. Following the death in 1708 of the tenth guru, Gobind Singh, the function of the guru passed to the Sikh scripture—the sacred writings—and to the community. For this reason, the Adi Granth (the Sikh scripture) is particularly respected by the community.

In the north of India in Nanak's day, a popular form of religion among ordinary people was the worship of a God of grace. This God of grace existed throughout all creation but never took the form of a person or an idol. The worship of this God was the Sant Tradition, and Nanak provided in his teachings its clearest statement. According to Nanak's teachings, the presence of God is known through the *nam*—the divine Name. This Name appears in the beauty and order of the world around us. It is one's duty to practice **meditation** focused on this holy Name. This may be done by repeating a particular word or **mantra**, singing hymns, or silent meditation. In meditating on this divine Name, one grows ever closer to God and eventually achieves a condition of perfect union with God. This union ends the cycle of the **transmigration** of the soul at death into a new body.

Those who accepted these teachings from Nanak were the first Sikhs. A line of gurus succeeded Nanak after his death. It was believed that the same divine spirit in Nanak inhabited each of the gurus who came after him. The first four gurus after Nanak continued his teachings concerning the divine Name. And from 1603 to 1604, Arjan, the fifth guru, collected the hymns of the four gurus and his own hymns into a sacred writing. He also added to it the works of other members of the Sant Tradition. During the time of the sixth guru, Hargobind, the community attracted the attention of the Moghuls, who were then the rulers of northern India. By this time the Sikh community had grown large, and the Moghuls were becoming suspicious of its increasing numbers. This danger seemed to pass away, but it returned at the time of the ninth guru, Tegh Behadur, who was executed by the Moghuls in 1675.

The Foundation of the Khalsa

In 1699, Tegh Behadur's son and successor, Gobind Singh, started the Khalsa. The Khalsa was a new order that loyal Sikhs were asked to join. Membership in the Khalsa was through an initiation ceremony and a lifelong vow to keep certain outward symbols. Such vows and symbols included not cutting one's hair. The Khalsa continued to emphasize the importance of meditating on the divine Name. But in place of the strictly inward faith taught by Guru Nanak, the tenth guru created an organization that proclaimed the identity of his followers to all.

The foundation of the Khalsa was very important, because it laid down a clearly stated code for its members. This code was known as the *Rahit*. According to tradition, the tenth guru made known all that the modern Khalsa observes today. In fact, many of the individual items of the *Rahit* can be traced to experiences that follow the actual foundation of the Khalsa. The essential nature of the Khalsa, however, remains unaffected. Gobind Singh, the tenth guru, summoned loyal Sikhs to join his Khalsa. The Khalsa Sikh was to be known by certain outward features. These included the obligation to bear arms and to keep one's hair uncut.

Ranjit Singh, the Singh Sabha, and Modern History

The eighteenth century in the Punjab region of India was a turbulent time. It was followed by an orderly and settled period during the early nineteenth century. A strong government was introduced under Maharaja Ranjit Singh, who became ruler of the central Punjab in 1801. During the next twenty-five years, the boundaries of the Punjab were enlarged in three directions. In the southeast Punjab, where British soldiers advanced against Ranjit Singh, the border was drawn along the Satluj river. This left many Sikhs in British territory or in the territory of their client states. Amritsar was not the capital city, but it was confirmed as the main religious center. Ranjit Singh covered the two upper stories of its main temple in gold, making it into the famous Golden Temple.

The death of Ranjit Singh in 1839 marked the beginning of a steep decline in the history and fortunes of the Sikhs. In 1849, following two wars, the British annexed the Punjab. In 1873 the Singh Sabha (Singh Society) was founded, and under its influence, the Sikh community was revived and reshaped. But in 1920 the Singh Sabha was taken over by the more radical Akali movement. This movement was dedicated to the liberation of the *gurdwaras*—the temples. With the partition of India in 1947 into a Hindu and a Muslim state, the Punjab was divided, and much of its territory went to the new nation of Pakistan. As a result, the Sikhs who were now in a mostly Muslim Pakistan moved across to the Indian area. Since 1947, many Sikhs have claimed greater independence for the Punjab. In fact, the Indian army assault on the Golden Temple in 1984 led to demands by many Sikhs for Khalistan, a completely independent state. But by the 1990s, these demands were for the most part dropped.

▲ The Golden Temple or *Darbar Sahib* (Punjabi for "divine court") is the main house of worship for Sikhs in India. It is the place where Sikhs from around the world come on pilgrimage. The temple is a marble structure with a huge dome covered in gold leaf. The original Golden Temple was completed in 1603; the rebuilt temple dates to the early nineteenth century.

In the twentieth century, Sikhs began to migrate to the United States. The Sikh Yogi Bhajan from New Delhi began teaching in Los Angeles in 1968. With his American following, he founded the Health-Happy-Holy Organization in 1969. There are an estimated ten thousand Westerners who have converted to the Sikh religion in Europe and in North America. Sikh men and women wear turbans and usually dress in the styles of the East, often in flowing white robes.

The Singh Sabha and the *Rahit*

The main concern of the Singh Sabha reformers was to show that Sikhs had an entirely distinct faith and that, in particular, they should not be confused with the Hindus. At this time, special concern focused on how a Sikh should behave.

This question concerning Sikh behavior required a restatement of the Sikh code, the *Rahit*. According to tradition, Guru Gobind Singh had made the *Rahit* known in all its details. But by the late nineteenth century, it had become impossible to determine which were his exact words. The *Rahit* had been recorded for Sikhs in a number of *Rahit* manuals, none of which was entirely satisfactory. Those who had been present at the signing of the Khalsa in 1699 would know what was required of them, as would those who associated with the tenth guru at the time of his death in 1708. But most of the eighteenth century was marked by warfare and persecution. Sikhs then had little time to record the *Rahit* that had been given to them. Sikhs believed that ignorant or mischievous people might have corrupted the *Rahit* and so the *Rahit* manuals could be trusted only after someone removed those portions that misled readers and restored those parts that had been lost.

▶ A Sikh priest leads from the Adi Granth, the sacred book of the Sikhs.

The Singh Sabha leaders at first made unsuccessful attempts to produce an authentic *Rahit* manual. Eventually an acceptable version, *Sikh Rahit Maryada*, came out in 1950. Unlike other religions, the Sikhs have no clergy, and so the publication of a text of such authority was truly important. But the question of orthodoxy—how to observe the faith—remains. *Sikh Rahit Maryada* represents the Khalsa version of orthodoxy—that one's hair should not be cut. There is no doubt that since the days of the Singh Sabha, this style has dominated. Yet there are some Sikhs who do not observe this. They hold the gurus and sacred writings holy but also choose to cut their hair. Even though they do not observe the *Rahit*, they still insist that they are Sikhs. It is here that Sikh identity becomes difficult to define and, with it, the whole question of what makes up Sikhism.

Khalsa Regulations

Members of the Khalsa are identified by what are called the Five *K*s (from the Punjabi words for them). These are uncut hair, a comb, a steel wristband, a sword or dagger, and shorts. Smoking and drinking alcohol are banned, although drinking does occur, and only the ban on smoking is strictly kept. Khalsa Sikhs insist on carrying a sword, something that contributes to their reputation for violence. This reputation is greatly exaggerated. In reality, the Sikh should use his sword only defensively, only when the cause is just, and only when all other methods have failed.

In Sikhism, the key term when discussing ethical and moral issues is *seva*, or service. Little guidance is given regarding health, disease, and the environment, other than the most general rules. The objective of Sikhism is simply a life of personal righteousness. However, this idea of doing what is right is not really spelled out. Service is mostly considered a duty toward the temple, and it consists of performing obligations for the guru on the temple's holy ground. These duties include service in the free dining room that all temples are required to maintain. This service symbolizes the idea that all people are equal. It is also interpreted to show genuine concern for the needs and welfare of others.

In general, Sikhs see themselves as different from other faiths, particularly from all forms of Hindu tradition. This is the case with funerals, which involve a simple rite. In the Sikh tradition, cremation follows death, but all who come are required to restrict their mourning. The dead body is dressed in clean clothes, complete with the Five *K*s, and the ceremony surrounding the burning of the corpse is carried out while hymns are sung.

Although in the Sikh religion each individual Sikh is expected to live a worthy life, matters of bioethical concern are not really spelled out. There are two exceptions, and Sikhs are left to determine them in the light of their religious faith. One exception is that female infants should not be killed. The second exception is that initiated Khalsa members should not eat from the same dish as an uninitiated Sikh or as one who has given up his faith. All other issues, such as **abortion**, birth control, suicide, and **euthanasia**, are left to the individual or the family to decide.

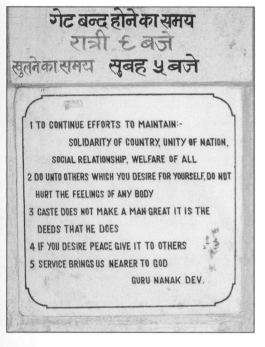

▲ An inscription bearing the five *K*s of Guru Nanak's teachings at Kulu, India.

abortion the deliberate or spontaneous early ending of a pregnancy, resulting in the death of a fetus

euthanasia mercy killing

TAOISM

holistic concerned with the whole body, not just a part of it; with the whole person, both mind and body; and even with the person within his or her immediate environment

harmony an orderly arrangement of elements that achieves a state of balance or tranquillity

balance a stability achieved by an equalizing of opposing forces; homeostasis

LAO-TZE

The Obscure Sage

▲ Portrait of the Chinese philosopher Lao-tzu (c.604–531 B.C.E.). The robed founder of Taoism rides on a bull.

spiritual relating to the nonphysical intelligence or feeling part of a person; having to do with sacred matters or religious values

Taoism can be viewed as a philosophy about nature and human life, described in ancient texts like the *Tao te ching*, written by Lao-tzu, and the writings of Chuang-tzu. Others see it as a religious tradition that emerged around the second century C.E. whose importance has lasted into the twentieth century.

Unlike many Western religions, Taoism does not have a coherent structure of concepts, so it is not easy to define a clear "Taoist position" on ethics or any other issues. But most Taoists share certain assumptions and concerns. Taoism is a **holistic** worldview, with related moral and religious values and practices. Taoists believe that all things in the universe—rocks and rivers and stars and people—are interrelated parts of a greater whole. In this view, something that affects one part of the universe affects every other part.

Taoism sees human reality as just a small part of a much deeper and greater reality. Western religions consider humans to be the most important part of creation. In Taoist thought, humans are only parts of the greater universe, neither more nor less important than animals, trees, stones, or the weather. Taoists believe that the universe is harmonious, and they try to become aligned with this **harmony**, as individuals and within society as a whole.

The Taoist Heritage

In ancient texts written by the Chinese philosophers Lao-tzu and Chuang-tzu, oneness with the universe is described as a return to the natural rhythm or flow of life. This rhythm or flow is called the Tao. The word Tao also refers to the source of the universe, the patterns or cycles of the universe, and the natural force that keeps everything in the universe in **balance**.

Harmony. An image often used to illustrate Taoism is of a farmer who is dependent on the natural harmony and cycles of nature to survive. The farmer tries to understand and work with the flow of the seasons and the cycles of nature. If he plants seeds in the winter, going against the natural cycles of nature, his crop will fail. If he plants when the earth and the seasons are ready, his crop will succeed.

Balance. In the Lao-Chuang interpretation of Taoism, based on the writings of Lao-tzu and Chuang-tzu, all the world's problems, such as war, are caused when humans lose a balanced view of reality and move away from the harmony of the Tao. The goal of Lao-Chuang Taoists is to regain a balanced view and realign themselves with the flow of the natural world, its forces, and its processes.

The first Lao-Chuang texts are vague about how people are supposed to do this, but later texts of Lao-Chuang Taoism, such as the *Kuan-tzu* and the *Huai-nan-tzu*, are more detailed. They suggest specific **spiritual** and physical practices that can be used to bring a person back into harmony with the natural forces of the universe.

Another important difference from Western thought is that Taoism does not separate body and mind or matter and spirit. These are considered aspects of the Tao and are not thought of as opposites, or separate from each other, as they are in the West.

The Evolution of the Taoist Religion

The Taoist religion can be seen as a river formed by the joining of many streams. Its origins go back to the Han Dynasty, which lasted from 206 B.C.E. to 221 C.E. Chinese intellectuals, such as the Confucian philosopher Tung Chung-shu, were then trying to form a complete explanation of the world and how it worked. Based on this explanation, advisers to the Chinese emperor wrote a series of **sacred** texts, including the *T'ai-p'ing ching*. It is thought to be the first Taoist scripture.

The first Taoist scripture. According to the *T'ai-p'ing ching*, ancient rulers had maintained "Grand Tranquillity" by following the principle of *wu-wei*, "nonaction." This principle involves avoiding intentional, purposeful action, and trusting instead in the world's natural order—the Tao. Taoists believe that if the world is left alone, it returns to harmony and stays that way. When humans interfere with the Tao and try to cause events to happen, they disrupt the harmony of nature, because they do not understand it.

According to the *T'ai-p'ing ching*, when later rulers tried to control events too much, they disrupted the "Grand Tranquillity." People must return to the Tao by looking within themselves, the scripture says. It gives specific directions for how to find union with the Tao. These directions include moral guidelines, as well as recommendations for increasing health and length of life through practices like controlled breathing, medicine, **acupuncture**, and music therapy. In this way the *T'ai-p'ing ching* gives people practical advice on how to regain harmony with the natural flow of the universe.

New influences: Shang-ch'ing. When northern China was invaded by nomadic people from the steppes to the north in the fourth century C.E., Taoist leaders fled south. In the south of China, they found a rich religious culture that focused on perfecting the individual through **ritual**. Unlike Taoism, the religion of southern China had little interest in making a perfect society. It focused almost completely on the individual.

The fourth century was a time of rich contact between different religious traditions in China. Two revelations from divine beings were thought to have occurred then. The first, called the *Shang-ch'ing*, the "Supreme Purity" revelation, was said to have been received from angelic beings called "Perfected Ones," who lived in distant heavens. The Perfected Ones revealed methods that people could use to ascend to the heavens. One key method was **meditation**.

The integration of alchemy. Shang-ch'ing Taoism also took on the older southern Chinese interest in perfecting the person

---

sacred dedicated or set apart for the worship of God or a god;

> "One may know the world without
> going out of doors.
> One may see the way of Heaven
> without looking through the
> windows.
> The further one goes, the less one
> knows.
> Therefore the sage knows without
> going about,
> Understands without seeing,
> And accomplishes without any action."
>
> —*Tao te ching*

acupuncture a method in traditional Chinese medicine of treating illness or relieving pain by puncturing specific points on the surface of the skin of the body

ritual having to do with a traditional ceremony or the established form of words of such a ceremony

meditation spiritual introspection or the contemplation of a religious truth, often practiced in order to reach a higher state of understanding or realization

alchemy an ancient philosophy or science that seeks to change one substance into another, to find a cure for all disease, or to make it possible to live forever

> "To keep the body in good health is a duty, for otherwise we shall not be able to trim the lamp of wisdom, and keep our mind strong and clear. Water surrounds the lotus flower, but does not wet its petals."
>
> —Buddha, *The Dhammapada*

liturgical related to the use of rites prescribed for public worship, or the use of customary ideas, phrases, or observances

through **alchemy**. Alchemy, which also became popular in Europe in later centuries, was a process of spiritual transformation that was described by using chemical symbols and terms.

Many people mistakenly believe that alchemy is a "typical" element of religious Taoism, but it actually developed independently. Scientific chemistry was not yet understood, and people saw chemical transformations as magical and mysterious. Some Shang-ch'ing Taoists became interested in alchemy because they believed that just as matter could be changed and refined through chemical reactions, so a person's spiritual state could be elevated and refined. They believed that a refined, perfected person might be able to live forever or ascend to the heavens.

Perfecting the self. Both the alchemical tradition and Taoism were concerned with perfecting the self. This idea was based on the assumption that there is no difference between people's physical body and their spiritual self. If the body can be purified by using alchemy and other ways to health and long life, then the moral and spiritual self can be refined at the same time.

Ling-pao. The Shang-ch'ing revelations were immediately followed by a very different set of revelations, known as the *Ling-pao* ("Numinous Treasure"). Ling-pao Taoism is different from other forms of Taoism because it includes elements of another religion, Mahayana Buddhism. It also involves a greater interest in human society and the community.

In the fifth century, the Ling-pao tradition was redrawn by Lu Hsiu-ching, who redesigned its rituals and wrote new ones that continue to influence Taoist practices today. A central liturgy is the *chiao*, a long series of rituals meant to renew the local community by reconnecting it with the harmony of heaven.

"**Liturgical** Taoism" soon became a major part of life in all parts of Chinese society. Chinese emperors tried to make themselves seem more important by associating with Taoist teachers and by having them perform rituals for the sake of the government and society. During the T'ang dynasty (618–906), cultural leaders in every field became associated with Taoist teachers and were deeply influenced by Taoist religious, artistic, and literary traditions.

In the twelfth and thirteenth centuries, China was again invaded by northern peoples, and the social importance of liturgical Taoism changed drastically. Because the new foreign rulers thought that religious organizations were fostering rebellion, Chinese people who wanted to advance socially or politically stayed away from religious groups. In late imperial China, liturgical Taoism became separated from the wealthy, powerful, and educated people in society and became a religion of the poorer and less-educated people.

Meditation as a way to perfect the self. Modern Taoism has kept the idea that a person can become spiritually perfected through meditation. Earlier Taoist meditation took different forms.

Buddha in Meditation, from the caves at Dunhuang, China. Meditation is one of the ways to spiritual perfection in Buddhism as well as in Taoism.

From the eleventh century, most Taoist meditation was described in chemical terms, as a kind of "inner alchemy." Using terms from ancient Lao-Chuang texts, "inner alchemy" aims at perfecting the self through cultivating *shen*, "spirit," and *ch'i*, "life force." These practices were important in *Ch'üan-chen* ("Complete Perfection") Taoism, a monastic movement founded in the twelfth century. Ch'üan-chen institutions continued into the twentieth century, and so did some of their teachings on perfecting the self through meditation.

The Ethical Dimensions of Taoism

Many accounts of Taoism lead one to question whether there is such a thing as a Taoist ethic. Some people believe that Taoism focuses too much on the individual and not enough on society. But all the different Taoist teachings and traditions really do have ethical guidelines for individuals, and most teach social values as well.

Promoting harmony. These ethical and social values were not borrowed from Confucianism or Buddhism but are a natural part of the Taoist worldview. They come from the ancient heritage of the *Tao te ching* and the *T'ai-p'ing ching*. The Taoist worldview encourages individuals and groups to do things that promote the healthy connections between the individual, society, nature, and universe.

The Taoist view of personal identity and human values is very different from another Chinese philosophy, Confucianism. Confucians believe that humans are by nature different from and better

The Old Man and the Horse

Taoism emphasizes the importance of accepting the course of life, rather than actively trying to shape the future. This is best illustrated by the Taoist story "The Old Man and the Horse."

Once in ancient China there was an old man who had a wonderful black horse. The neighbors told him how fortunate he was, to which he responded, "How do you know?" One day the horse ran away. The neighbors told the old man how unfortunate he was, to which he responded, "How do you know?" The horse returned with two wild stallions by its side. The neighbors told the old man how fortunate he was, to which the old man responded, "How do you know?" The old man's only son went riding on a stallion and broke his leg. The neighbors told the old man how unfortunate he was, to which he responded, "How do you know?" At that time all the young men of China were being forced into armies of laborers to build the Great Wall of China. It is said that for every brick placed in the wall, a young man died of exhaustion. The old man's son was allowed to remain at home. The neighbors told the old man how fortunate he was, to which he responded, "How do you know?"

▲ The three sages: Buddha, Confucius, and Lao-tzu.

"Things in their original nature are curved without the help of arcs, straight without lines, round without compasses, and rectangular without squares; they are joined together without glue and hold together without cords. In this manner, all things live and grow from an inner urge and none can tell how they come to do so. They all have a place in the scheme of things and none can tell how they come to have their proper place. From time immemorial this has always been so, and it may not be tampered with."

—Chuang-tzu

than all other forms of life. This is because humans, unlike animals, are socially aware, and also because they are capable of knowing right from wrong.

A universal perspective. Taoism is unlike Confucianism because it believes that humans are not special as a result of their being different from other animals or from the rest of nature. Taoists believe humans are valuable because they are connected to and part of nature and the overall harmony of the universe. Taoists constantly try to see things from a higher, wider perspective. If we go back to the example of the farmer, he might dislike having a rainy day, but if he takes a longer view, he knows that rain is part of the flow of nature, the cycle of the seasons, and is necessary for both him and the crops. Taoists try to see not just beyond the narrow concerns of the individual, but also beyond the narrow concerns of society. Taoists do not ignore social concerns but try to rise above them.

Leaving nature alone. Even though Taoism stresses living in harmony with nature, Taoists tend not to be environmental activists. This is because of the desire to rise above conflicts and to become detached from everyday concerns. The *Tao te ching* sees nature as more powerful than any human action; so if people simply take no action, nature will balance itself. Of course, "not taking action" does not refer only to not taking action to help the environment. It also means that humans would stop doing anything that affects the environment in any way, such as polluting and cutting down forests. If we simply leave nature alone, Taoists believe, it will maintain its own balance.

Pursuing personal improvement. Taoists insist that we must look at ourselves and try to understand and become part of the deepest realities of the universe through *lien*, a process of personal refinement. In some variations of Lao-Chuang Taoism, this process is so abstract that it seems to involve only a change in perception; through it, a person learns to reject what seems obvious in order to see things more deeply. But most later Taoists see the process of improvement of self as a larger task, transforming one's physical as well as spiritual reality.

Human lives are supposed to reflect the workings of the Tao. The Tao is very different from Western images of God as creator, father, ruler, or judge. It is not a god or an external authority; it is not a being that has a moral right to control or intervene in the lives of people. Also different from Western ideas is that the *Tao te ching* values "feminine" behaviors like yielding and passivity, as opposed to "masculine" behaviors like assertion, intervention, or control. So Taoists are not likely to take control and "play God," even in medicine, government, or law.

Imbalance as the source of disease. Although Taoism did not create the ideals of balance and harmony, it embraced them more than any other tradition has. A basic Taoist belief, which applies to everything in life, is that disorder results from imbalance.

The imbalance may be physical or spiritual, individual or social. If someone is physically ill, Taoists see this as showing a "biospiritual" imbalance within the individual. The sickness can be physical or spiritual, because Chinese medicine does not distinguish between them.

In Chinese medicine, disease is believed to be caused by a disruption of the natural life force, *ch'i*. The concept of *ch'i* can be confusing, even to the Chinese, because it also can be seen as a spiritual force. Uneducated people sometimes believe that people get sick because they have done something bad. Some Taoists, wanting to involve the common people, even encourage this thinking by including such ideas in their writings and practices.

Restoring harmony is never something that each person could do alone, either in Taoism or in Confucianism. Just as a physical disorder is believed to result from an imbalance of body and spirit, so sociopolitical disorder is believed to result from a much bigger physical and spiritual imbalance in the society. Taoists and Confucians in classical times and in the later imperial period felt that they should work to fix the imbalance, and restore *T'ai-p'ing*, or "Grand Tranquillity." *T'ai-p'ing* referred to an orderly society, both in spiritual terms and in terms of the local community.

Gender issues. The Taoist concern with balance and harmony extends to participation in religious activities. The *Tao te ching* admires "feminine" traits, and in the early Taoist community, women participated equally with men. It is not clear how often women led rituals or had the same official functions as men, but medieval texts describe women's spirituality as very similar to men's. After the twelfth century, when China was again invaded by northern peoples, women's roles changed throughout Chinese society. They had fewer opportunities in general, including within Taoism.

Taoism and sexuality. Taoists never set out any clear or specific sexual ethic. Apart from a few Confucian moralists, few Chinese thought that there were moral problems associated with sexuality, and most thought it was a valuable part of human life. Some Taoists believed that the forces of reproduction were the most visible signs of the natural creative forces of the universe.

Taoist attitudes toward death. People disagree about the Taoist attitude toward death. Some people, looking at the traditional Taoist interest in longevity, say that religious Taoists try to develop practices to avoid death, but that the earlier Lao-Chuang Taoists encouraged acceptance of death, because it was a natural conclusion to the cycle of life. There is evidence to support this, but there are also passages in the *Tao te ching* and other Lao-Chuang texts that suggest that avoiding death and becoming immortal are possible and desirable.

Taoists of all periods would be puzzled by how Western medicine emphasizes preventing death as the most important concern in human activity. To Taoists, a person's existence goes far beyond the

---

Disease being the result of moral faults could be cured by public confession, repentance, and punishment. Good physical and spiritual health and long life were maintained by positive measures, such as meditation, and by various exercises in dietary and sexual hygiene.

Failure to attain Grand Tranquillity was caused by the existence of obstacles to the circulation of *ch'i*, or "essences." In relations between the sexes, sexual abstinence was condemned because it upset the interaction of *yin* (female) and *yang* (male) essences. The killing of infant girls was severely condemned because the practice caused an imbalance in the strength of *yin* and *yang*.

see also

Sex and Gender: Women

"Let life ripen and then fall.
Will is not the way at all:
Deny the way of life and you are dead."

—Lao-tzu

biological processes of the body, so extending the life of the body for its own sake would not make sense. Although the Taoist goal is to always be in harmony with the flow of the universe, Taoists do not see "life" in purely physical and biological terms. The Western interpretation of "life" as purely biological seems narrow and wrong to a Taoist.

Even though it is hard to clarify the Taoist attitudes toward death, it is clear that a search for immortality simply for the purpose of avoiding death was never a Taoist goal. Instead, Taoists always try to reach a state of harmony or perfection. This state of perfection or ultimate harmony has sometimes been called "deathless" or "immortal," but in many cases this is only a metaphor and does not actually refer to avoiding physical death.

Summary. Although some Taoist writings present moral rules, Taoism never developed any specifically ethical code. The idea of ethical guidelines as separate from other guidelines for living does not make sense within Taoism. Since the Taoist worldview does not include a divine Lawgiver, Taoists never developed an ethic that was seen as obedience to divine authority.

If Taoism does have an ethic, it is an ethic of **virtue**. In fact, the term *te* in the title of the *Tao te ching* is usually translated as "virtue." But the Taoist ethic is very different from the virtue ethic developed by Confucians.

To understand the moral aspects of Taoism, one must understand the concept of the Tao. The Tao goes beyond physical reality but is present in all of it. It is divine, but exists in people and in all of nature. Most important, since the Tao never acts by design, Lao-Chuang Taoists do not believe that any good could come from a person who consciously thinks about options and then deliberately decides what to do. They believe this kind of purposeful thought and ethical choice blinds people to the natural course of action, which follows the Tao. The natural course of action is what people do when they are spontaneous, and do not impose their will or ego on a situation. Taoists believe that people should live like rivers, flowing naturally, in harmony with all other forces of the universe.

Taoism has ethical dimensions, but they are subtle. Since the original Taoists never believed in making absolute rules about what people should or should not do, one must look at the entire Taoist tradition to see its ethics. Living a good Taoist life means living in a way that restores and keeps the universe's natural harmony. Taoists are dedicated to improving themselves, both physically and spiritually. As the individual improves, so does society. Taoists believe that they are parts of a harmonious universe and are responsible for maintaining the harmony, health, and wholeness of the individual, society, nature, and the universe.

see also

Death and Dying: Death

virtue moral excellence; goodness; behaving according to ethical or moral principles

---

"The best (man) is like water.

Water is good; it benefits all things and does not compete with them.

It dwells in (lowly) places that all disdain.

This is why it is so near to Tao."

—Lao-tzu

HUMAN RESEARCH

This entry consists of three articles explaining various aspects of this topic:

HISTORICAL ASPECTS Throughout the history of Western civilization, healers have tested the effectiveness of drugs or treatments by using them on other people, and seeing whether they work. But there are few records of how often these physicians did experiments, what they experimented with, and who their experimental subjects were. The most frequently mentioned cases were testing poisons on prisoners who were sentenced to die. There are few records of other research on human beings, and we are not sure how much research was done, and what treatments were involved.

We do know that people experimented on other people often enough to discuss whether this research was right or wrong, and what ethical rules experiments should follow. Moses Maimonides (1135–1204), a Jewish physician and philosopher, told other doctors that they should always treat patients as ends in themselves, not as means for learning new truths. This means that the patient's welfare is more important than whatever new medical facts one could discover by using the patient in an experiment.

Edward Jenner and Smallpox

Edward Jenner (1749–1823) was the first person whose experiments on human subjects made a significant impact on medical practice. In his time, a contagious disease called smallpox was killing or disfiguring many people in Europe. Jenner noticed that farm workers sometimes caught a similar, but much milder, disease called cowpox from cows or swinepox from pigs. He discovered that these people were then immune to smallpox. They did not get smallpox when they were exposed to it.

Both cowpox and swinepox caused people to break out with sores all over their bodies. Jenner decided to take material from the sores of a patient and inject it into another person. Then he would see if the second person became immune to smallpox. This deliberate transfer of material from one person to another was called inoculation or vaccination. In November 1789, Jenner inoculated his son, who was about a year old, with swinepox. But when he

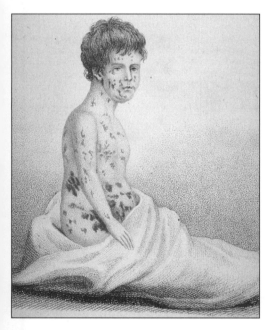

The sores of smallpox are vividly depicted on this early nineteenth–century image.

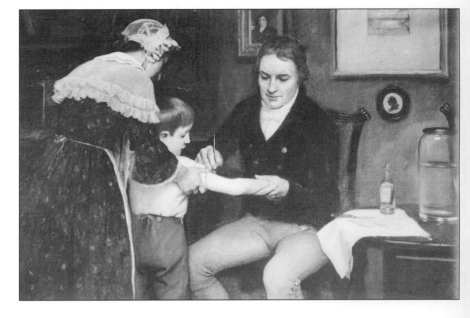

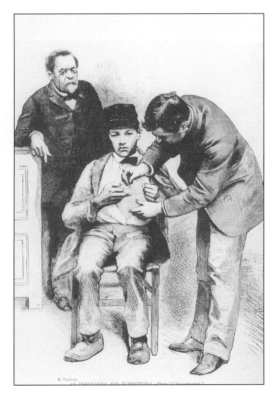

Edward Jenner vaccinates a young boy.

exposed his son to smallpox, the boy caught the disease. The inoculation had not worked.

Several months later, he tried again with another boy, who was about eight years old. This time he used cowpox material for the inoculation. A week later, Jenner injected the boy with smallpox, and nothing happened. The boy was now immune to smallpox.

We do not know if these boys were willing or unwilling to be in the experiment, or how much they understood of it. Jenner was well known and respected, and fortunately the experiment was beneficial to the second boy.

Louis Pasteur and Rabies

The most brilliant medical experiments of the nineteenth century were done by Louis Pasteur (1822–1895), who was sensitive to the ethics of his investigations. Even while he did research on animals to find an antidote to rabies, a fatal disease carried by dogs, he worried about what he would do when it was necessary to test the product on human beings. "I have already several cases of dogs immunized after rabic bites," he wrote in 1884. "I take two dogs: I have them bitten by a mad dog. I vaccinate the one and I leave the other without treatment. The latter dies of rabies: the former withstands it." But, he continued, "I have not yet dared to attempt anything on man, in spite of my confidence in the result. . . . I must wait first till I have got a whole crowd of successful results on animals. . . . But, however I should multiply my cases of protection of dogs, I think that my hand will shake when I have to go on to man."

The fateful moment came nine months later, when a mother came to him for help. She begged him to help her son, Joseph Meister, who had been severely bitten by a rabid dog. Pasteur struggled

Louis Pasteur observes as a young boy receives an inoculation for hydrophobia (rabies).

Claude Bernard (1813–1878), French physiologist, believed that experiments on humans should be done as long as the test subjects were not hurt.

with the decision of whether or not to use his vaccine on a human being for the first time. He asked two other doctors for their advice, and had them examine the boy. They urged him to go ahead. Because of their urging and also since the boy would definitely die without his help, Pasteur decided to go ahead. With great anxiety, he gave twelve inoculations to the boy. It took weeks before he could relax and become confident that his treatment did work. Joseph Meister was cured and would remain healthy.

Claude Bernard and the Ethics of Research

Claude Bernard (1813–1878), who was a professor of medicine at the College of France, not only did important research in physiology but also wrote about the methods and ethics of experimenting on people. "Morals do not forbid making experiments on one's neighbor or one's self," he wrote in 1865. But, he said, one should never perform an experiment that might hurt someone, even if it would lead to scientific advancements that would help many other people.

Bernard allowed some exceptions to this rule. He believed it was allowable to experiment on patients who were dying and on criminals about to be executed, on the grounds that "They involve no suffering of harm to the subject of the experiment." But he made clear that scientific progress did not justify violating the well-being of any individual.

> The French doctor Claude Bernard wrote *An Introduction to the Study of Experimental Medicine* in 1865. He stated: "Among the experiments that may be tried on man, those that can only harm are forbidden, those that are innocent are permissible, and those that do good are obligatory."

Walter Reed and Yellow Fever

As medicine became more scientific, some researchers violated ethics in experimentation. Medical progress—rather than the subject's welfare—was the goal of their research. Probably the most famous experiment in this gray area was the work of the army doctor Walter Reed (1851–1902). When Reed began his experiments, people knew that mosquitoes seemed to help spread yellow fever, but no one knew exactly how. To understand more about how the disease was spread, Reed began a series of human experiments while stationed in Cuba. Like many other researchers, he used the members of his research team as the first subjects.

The first case. After a time he realized that he would need more subjects. Soon, a soldier stopped by. "You still fooling with

▲ Walter Reed (1851–1902), American Army surgeon, developed the technique that resulted in the first case of yellow fever produced experimentally.

mosquitoes?" the soldier asked one of the doctors. "Yes," the doctor replied. "Will you take a bite?" "Sure, I ain't scared of 'em," the soldier said. So the soldier was bitten by an infected mosquito, and he soon had the first case of yellow fever to be produced experimentally.

Contracts with test subjects. After one of Reed's fellow investigators, Jesse William Lazear, died of yellow fever from purposeful bites, Reed and the other members of the research team decided it was too dangerous to test the infection on themselves. Instead, Reed asked American soldiers to volunteer, and some did. He also recruited Spanish workers and drew up a contract with them. The contract stated that the volunteer knew that if he got yellow fever, it might endanger his life to a certain extent. The contract also stated that the volunteer knew that he was likely to get yellow fever anyway, since the experiments were taking place in a tropical area. If he got the disease intentionally, under the care of the experimenters, he would receive "the greatest care and most skillful medical service."

Volunteers received $100 in gold, and those who actually got yellow fever received a bonus of another $100. If they died, the money would go to their heirs. Although twenty-five volunteers became ill, none died.

Reed's contract with the subjects was a step along the way to more formal arrangements with human subjects, complete with money or other enticements to be involved in a hazardous experiment. But Reed's contract was also misleading. In subtle ways, it distorted the risks and benefits of the research. His contract said that yellow fever was a danger to life "only to a certain extent," and did not mention that the disease might be fatal. Also, the contract made it seem as if the subject would probably get yellow fever anyway, which was an exaggeration aimed at getting volunteers.

Human Research and the War Against Disease

In the twenty years after the end of World War II, there was an extraordinary explosion of human experimentation in medical research. During the war, thousands of sick and wounded American soldiers needed help and the war seemed to justify using subjects without getting their complete consent. The benefits that could come to many people by the possible suffering of some subjects seemed to make the experiments worthwhile.

Long after peace had returned, experimenters in the United States continued to work in this way. Instead of actual war, they were concerned with the Cold War with Soviet Russia, and the war against disease. These worries seemed to justify doing experiments based on the ethic of finding "the greatest good for the greatest number" instead of concern for the individual subject.

National Institutes of Health. The driving force in post–World War II research in the United States was the National

Institutes of Health (NIH). The NIH was created in 1930 as an outgrowth of the research laboratory of the U.S. Public Health Service. The NIH became important as it took over the work of an organization called the Committee on Medical Research. This committee had been formed during World War II and oversaw research on dysentery, influenza, malaria, wounds, venereal diseases, and physical hardships.

In 1945, the government gave the NIH $700,000 to spend on research. By 1955, the annual amount had climbed to $36 million, and by 1970, to $1.5 billion. This sum allowed the NIH to award 11,000 grants to experimenters, giving them money so they could conduct their research. About one-third of these research programs involved experiments on human beings. The NIH administered a research program at its own Clinical Center and also funded outside investigators.

The Clinical Center assured its subjects that it put their well-being first: "The welfare of the patient takes precedence over every other consideration" (NIH, 1953). In 1954, a Clinical Research Committee was established to develop principles and to deal with problems that might come up in research with normal, healthy volunteers. Still, the relationship between investigator and subject was very casual. The investigator could decide what information would be told to the subject, or if the subject would receive any information at all. Usually the researchers did not tell the subjects much, because they were afraid this would stop subjects from being part of the researcher's experiment. No formal rules applied to researchers who worked outside NIH but used its money for their investigations.

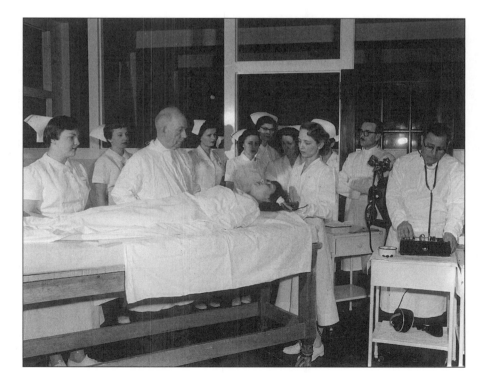

A volunteer nurse is about to undergo electric shock therapy at a hospital demonstration.

During the Holocaust (1933–1945) Germans killed as many as six million Jews in what is considered to be one of the most evil events of the twentieth century. Nazi doctors used Jews as guinea pigs for their medical experiments. Some of these Nazi doctors were given the death penalty at the Nuremberg trial of war criminals at the end of World War II.

Josef Mengele, one of the better known Nazi doctors, conducted experiments on inmates at the Auschwitz death camp. He escaped prosecution and fled to Paraguay, where he died in 1979.

Eva Moses-Kor is a Jewish victim of Nazi medical experimentation. In 1944, at the age of nine, she and her sister were taken from their village in Transylvania to Dr. Josef Mengele's laboratory. There they were injected with deadly germs and poisons. Eva managed to survive but her sister did not.

Lack of guidelines. The looseness of rules for the treatment of subjects shows how completely people were convinced at the time that science could find answers to the treatment of every disease, and that this advance of science was more important than almost anything else. This faith in science was so strong that the NIH did not establish guidelines to govern the research it paid for. By 1965, the program of grants for outside research was the largest source of research money for universities and medical schools. The NIH estimated that it supported between 1,500 and 2,000 research projects involving human research. Still, when the NIH gave money to investigators, it did not give them any rules for the ethical treatment of human subjects. The universities where the research was being done did not provide rules either.

In the early 1960s, only nine of fifty-two American departments of medicine had a set of rules for approving research that involved human subjects, and only five more said that they were in favor of having rules or wanted them in the future.

Nuremberg Code

People were shocked after the war by the horrible acts that Nazi doctors inflicted on human subjects (largely Jews and Gypsies), and they wanted more rules for ethical treatment of human subjects in experiments. The Nazi experiments included removing oxygen to learn how long a person could live without it, or deliberately infecting people with deadly bacteria or viruses in order to study the effects of drugs and vaccines. The horror that this inspired in people might have sparked Americans to insist on regulation of research, but this did not happen.

The Nuremberg Code was written in 1947 as a response to the Nazi atrocities. Its rules also applied to the medical research done in the United States. "The voluntary consent of the human subject is absolutely essential," the code said. "This means that the person involved should have legal capacity to give consent." But American researchers did not respect this principle.

The Nuremberg Code also stated that the research subject should be able to choose freely whether or not to be a subject. American researchers used prisoners as research subjects, and it is unlikely that the prisoners felt free to choose. The Nuremberg Code stated that human subjects should have enough knowledge and understanding of the experiment to make an informed choice about it, but some American researchers used mentally disabled people, who could not understand the experiment, as subjects.

Even though Americans were not following the Nuremberg Code, no one paid much attention to this until the early 1970s. Only a few articles in medical journals talked about the fact that the Nuremberg Code should apply to the ethics of medical experiments on humans in the United States. Perhaps people did not talk

The Declaration of Helsinki, adopted in 1964 by the World Medical Association, makes recommendations for conducting experiments on human subjects. The text, revised at following world conferences, features a general introduction, a list of basic principles, and recommendations for ethical treatment of subjects. Here are some of the Declaration's most important points:

- Biomedical research involving human subjects must conform to generally accepted scientific principles.
- The responsibility for the human subject must always rest with a medically qualified person and never rest on the subject of the research, even though the subject has given his or her consent.
- Concern for the interests of the subject must always prevail over the interests of science or society.
- In any research on human beings, each potential subject must be adequately informed of the aims, methods, anticipated benefits, and potential hazards of the study and the discomfort it may entail.
- In case of legal incompetence, informed consent should be obtained from the legal guardian in accordance with national legislation.

about it because they did not want to think they could be like the Nazis, or they did not want to think about what the Nazis had done. It is also likely that most experimenters did not think their experiments had anything in common with the horrible crimes the Nazi doctors committed. From their perspective, the Code did not apply to scientists in general. They thought it only applied to the Nazis. The guilty Nazi experimenters were seen as evil henchmen carrying out Adolf Hitler's orders, not as doctors.

Moral Progress

In 1964, the World Medical Association issued the Helsinki Declaration, which stated general principles for human experimenters. This document has since been revised four times. The declaration is modeled on the Nuremberg Code and requires investigators to be qualified and to have consent from the subjects. The 1975 revision recommended that an independent committee review the research to be sure it follows the guidelines.

Ethics in research. In 1966 Henry Beecher, a professor of anesthesia at Harvard Medical School, published an important article called "Ethics and Clinical Research" in the *New England Journal of Medicine*. This article described how twenty-two different researchers risked the health or life of their human subjects without telling them of the dangers or getting their permission. In one case, investigators purposefully did not give penicillin to soldiers who had streptococcal infections in order to test other methods for preventing complications. The men had no idea that they were part of an experiment or that they could get rheumatic fever, a dangerous disease. Twenty-five of them did get rheumatic fever. Beecher's conclusion was that "unethical or questionably ethical procedures are not uncommon" among researchers. Although he did not name the investigators in his article, he said that the ethically disturbing experiments were done in leading medical schools, university hospitals, private hospitals, governmental military departments, government institutes (including the National Institutes of Health), Veterans' Administration hospitals, and industry.

Unethical research. Two of the cases that Beecher described were important because they made the public aware and angry about unethical human research that was being done. In one case, researchers gave live hepatitis virus to mentally retarded children who lived in a New York state institution called Willowbrook. The researchers wanted to study the course of the disease and ways to create a vaccine against it. The other case involved doctors who wanted to study the body's immune response. They injected live cancer cells into twenty-two elderly and senile patients at the Brooklyn Jewish Chronic Disease hospital without telling them that the cells were cancerous.

▶ In 1997 President Bill Clinton apologized to a survivor of the Tuskegee syphilis experiment for the grossly unethical research that left African-American men with syphilis untreated.

In *Bad Blood: The Scandalous Story of the Tuskegee Experiment: When Government Doctors Played God and Science Went Mad* (1981), James H. Jones describes how from 1932 to 1972 the U.S. Public Health Service and private doctors deliberately withheld treatment from more than 400 African-American men who were suffering from syphilis. Many of the men died from the disease. None of the researchers were ever charged with criminal acts. In the 1990s the U.S. government finally paid settlement fees to the surviving patients or their families.

Another case that sparked fierce public and political reactions in the early 1970s was the Tuskegee research of the U.S. Public Health Service (PHS). Since the mid-1930s, its researchers had been visiting Macon County, Alabama, to examine, but not treat, a group of African Americans who were suffering from syphilis. The PHS could not give a good reason for not treating these people, but what really outraged the public was the fact that the PHS asked the draft board not to call these people to army service, because the army might treat them, cure them, and ruin the experiment. Even worse, in the early 1940s, once penicillin was discovered to be a good treatment for syphilis, the PHS had refused to provide good medical care and treatment, including penicillin, for these subjects.

Regulating Human Experimentation

The cases Beecher wrote about and others that were publicized in the press between 1966 and 1973 resulted in big changes in policies at NIH and the U.S. Food and Drug Administration (FDA). Both agencies were very sensitive to pressure from Congress because they were afraid that Congress would vote to take their money away. They also realized the conflict of interest in the researcher-doctor. Doctors work for the welfare of their subjects. But experimenters want knowledge more than anything, while their subjects want cure or well-being.

New guidelines. In February 1966, and in a revision in July 1966, the NIH, through the Public Health Service, publicized guidelines covering all human research that the government pays for. The revision of July 1966 stated that when a university or other institution received money from the NIH, it was responsible for getting the patients' consent and keeping records of this consent. It also said that a group of health-care professionals would have to review the project to be sure it met ethical guidelines. It defined the standards that this committee would follow: "This review must address itself to the rights and welfare of the individual, the methods used to obtain informed consent, and the risks and potential benefits of the investigation." For the first time, ethical decisions that used to be left to the conscience of the experimenter were watched over by a group of qualified and knowledgeable people.

Institutional Review Boards

The new set of guidelines did not interfere with experiments as much as some investigators feared. The guidelines were also not as protective of patients as some people wanted them to be. The core requirement was that a committee of professionals in the same field as the investigator would oversee and approve the experimenter's methods. This group was called the Institutional Review Board. Now investigators could not make their own decisions about whether an experiment was ethical but had to answer to fellow workers who would force them to uphold the government guidelines.

This set of guidelines for human research, written in 1966, accomplished what the Nuremberg Code had not. The guidelines made medical experimentation more public and showed the evils that could happen when ethical decisions were left up to the individual investigator.

A medical researcher at work on a microscope with video equipment at a table full of electronic devices.

Research on Drugs

In August 1966, the FDA issued a "Statement on Policy Concerning Consent for the Use of Investigational New Drugs on Humans." This statement distinguished between research that was a part of good and acceptable treatment (therapeutic) and research that was not (nontherapeutic). Like the Helsinki Declaration and other international codes, it prohibited all research that was not a part of treatment, unless the patients gave consent. When research involved "patients under treatment," and might improve their condition, the researcher had to obtain the patient's consent, except in the rare cases where the patient could not communicate or where the process of receiving consent would make the patient worse.

Advice on consent. Also, the FDA, unlike the NIH, was very clear about what "consent" meant. To give consent, the person had to be able to exercise choice and to have a "fair explanation" of the procedure, including understanding what the experiment was for and how long it would last. The patient also had to understand all the possible inconveniences and hazards of the experiment, what the methods would be, and that as part of the experiment, he or she might receive the treatment or might not receive it at all. The patient should also be told what other treatments were available.

Balance between researcher and subject. The FDA guidelines set a new stage in the balance of authority between researcher and subject. The insistence that the researcher obtain the subject's consent for all research that was not part of treatment would have eliminated many of the unethical World War II experiments as well as most of the cases Beecher wrote about. The FDA's definition of consent was much better than the vague NIH definition, and made the patient's consent much more significant.

Some parts of the process were still unclear. The FDA still confused research and treatment, and its rules governing investigations that were a part of treatment still gave a lot of power to the experimenter. But in general, authority was taken away from the individual researcher. A group of colleagues and the human subjects who were involved in the experiments received power to take part in making decisions.

Commissioning Ethics

The U.S. Congress became concerned about human experiments and medical ethics. In 1974, it created the National Commission for the Protection of Human Subjects. This commission recommended regulations to federal agencies to protect the rights and welfare of human research subjects.

The idea for this commission was supported by people's awareness of the awesome power of new medical technologies. Publicity about abusive experiments like the Tuskegee trials also helped convince Congress to create the commission.

The National Commission for the Protection of Human Subjects was composed of eleven members taken from "the general public and from individuals in the fields of medicine, law, ethics, theology, biological science, physical science, soil science, philosophy, humanities, health administration, government, and public affairs." The length of this list, and the rule that only five or fewer of the members of the commission could be researchers, showed how serious Congress was about bringing human research into public scrutiny.

Senator Edward M. Kennedy, who chaired the hearings that led to the creation of the commission, repeatedly stated that policy on medical experiments should come not just from doctors but also from "ethicists, the theologians, philosophers, and many other disciplines."

Although the National Commission was temporary and only made recommendations to the Secretary of Health with no power to enforce its recommendations, most of the policies it recommended became law. This tightened up the regulation of human experimentation. The National Commission agreed that institutional review boards should oversee research and recommended special protection for prisoners, mentally disabled people, and

The National Research Act of 1974 provided for the creation of the National Commission for the Protection of Human Subjects of Biomedical and Behavioral Research. The commission served from 1974 to 1978.

In 1978 the National Commission for the Protection of Human Subjects of Biomedical and Behavioral Research issued the Belmont Report: Ethical Principles and Guidelines for the Protection of Human Subjects of Research. The principles stressed in the report were: respect for persons, beneficence, and justice.

children. It recommended that an Ethical Advisory Board be established within the Department of Health and Human Services to deal with difficult cases.

This board was inaugurated in 1977 but expired in 1980, leaving a gap in the commission's plan to oversee research ethics. However, the Office for Protection from Research Risks at NIH watched over institutions to make sure they obeyed rules for research. The commission issued the Belmont Report, which was a statement of the ethical principles that should govern research. These principles were respect for autonomy, beneficence, and justice. The Belmont Report not only had an influence on research ethics but also on the newly developing discipline of bioethics.

UNETHICAL RESEARCH Whatever ethical framework we use, if we look at the history of human research, we can find many grim examples of violation of ethical principles.

Germany. During World War II, German researchers performed many experiments in concentration camps and elsewhere. Most of the victims of Nazi research were Jews, but they also included Gypsies, prisoners of war, political prisoners, and others. Nazi experimental atrocities included studies of quicker and more efficient ways to sterilize people so they could not have children. This involved radiation and castration of men and women without anesthetics. Some of the best-known cases involved causing death by freezing and investigating ways to prevent it. The Nazis wanted to do these tests because many of their pilots had died after crashing into the cold water of the North Sea. The tests included immersing prisoners in freezing water for long periods of time and watching to see what happened to them.

United States. In 1993, the public found out that the U.S. government had been doing experiments on the effects of radiation on humans beginning in 1945. In one study, conducted from 1945 to 1947, eighteen patients who were considered to be terminally ill were injected with high doses of plutonium to determine how long the radiation stays in the human body.

Because radiation and atomic energy were military secrets, the patients could not be told what was going on, so they could not give informed consent. Instead of telling the subjects that they would be injected with radioactive plutonium, the investigators told subjects that they would receive "a product."

Experiments on mentally handicapped teenagers in a Massachusetts institution involved feeding the subjects very small amounts of radioactive iron and calcium to see how the body absorbed them.

In reaction to these and other studies, orders were issued in 1993 to make public the government documents about U.S. service-

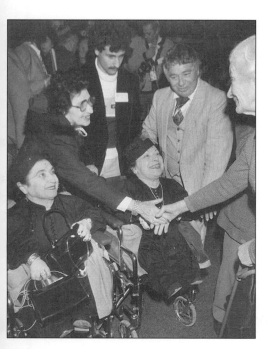

Dr. Ella Lingens, right, shakes hands with a relative of Sarah and Perla Qoici, seated in wheechairs. The twin sisters were victims of Nazi doctor Josef Mengele's experiments on dwarfs. Dr. Lingens, a Viennese doctor, was incarcerated at Auschwitz for aiding Jews. This picture was taken after Dr. Lingens testified against Mengele in a mock trial held in Jerusalem, Israel, in 1985.

men and others who were exposed to radiation from atomic-weapons testing after World War II. In 1994, President Bill Clinton appointed a panel to guide a federal investigation into the radiation studies.

Why this happens. Several themes emerge from the examples of unethical research that we know about. These studies are likely to be done using people who have little power or influence about what is done to them. If there is no public outcry, unethical research may continue for many years, even if people who read the scientific reports of the studies have all the facts they need to expose the unethical practices. The larger and more unethical studies are especially likely to be motivated by concerns about national security, and are often paid for by the military.

RESEARCH POLICY
Public policies that govern biomedical research will always conflict with the goals of the research. Sometimes this tension results in major controversies. In other situations, good public policy can reduce conflict.

Despite continuing tension, the public policies that both support and restrict biomedical research in the United States have produced a flowering of biological knowledge that is unique in human history. If investigators receive government funding for their work, they must follow the established rules and be accountable to the public for their actions.

Vulnerability to Coercion and Manipulation

Most people agree, at least in theory, on how to protect the rights and welfare of competent adults who are vulnerable to being manipulated or forced into being research subjects.

Need for review. Studies should be reviewed to make sure that subjects truly know what they are getting into, that they are truly volunteers, and that the risks of research do not affect one group more than another. This evaluation should be done by panels known as institutional review boards (IRBs). But the views of the IRBs may be different from the views of vulnerable subjects. One way to fix this is to assemble a group of people who might be subjects and ask them what they think and feel about the experiment.

Protection of subjects. Second, most people agree that in cases where the subjects are more vulnerable and the experiment is more risky, the protection of subjects should be tighter. If it is difficult to tell how voluntary a vulnerable subject's consent is, then the subject should have special legal protection. Because prisoners do not have their freedom, the U.S. government has special regulations about research on prisoners. If a study involves more than a small risk, or will not benefit the prisoners, researchers must prove that

In 1997 the National Institutes of Health (NIH) called a special meeting for the purpose of reviewing guidelines that rule research on people who are mentally ill or who have Alzheimer's disease. Special safeguards are needed so that these people will not be harmed in research.

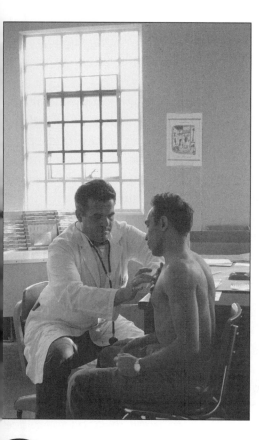

A doctor examines a prisoner at the Vacaville State Prison who is on a synthetic diet suitable for space travel.

there is a good reason to do the experiment. They must also provide special safeguards, get approval from experts, and receive authorization from the Secretary of Health and Human Services.

These restrictions make it difficult for researchers to give a good reason for experimenting on prisoners, because there is no disease or condition that affects only prisoners. But because prisoners have such a restrictive and unusual living situation, researchers could justify doing social or behavioral studies with them.

Avoid pressure. A third area of general agreement about protecting vulnerable competent adults from coercion or manipulation has to do with avoiding interference with people's right to decide what is best for them. But there is less agreement about how to do this. Some guidelines, such as the U.S. federal government's "Protection of Human Subjects," regard pregnant women as especially vulnerable to pressure and coercion. The guidelines instruct the IRBs that when some or all of the subjects are likely to be children, prisoners, pregnant women, mentally disabled persons, or economically or educationally disadvantaged persons, there must be extra safeguards to protect their rights and welfare.

Even when vulnerable people are competent, there are many disagreements about what restrictions on their choices are fair, promote their well-being, and respect their right to choose what is best for them. If there is too little protection, they may be exploited, but if there is too much, it may interfere with their freedom to choose for themselves. Before limiting the freedom of competent people, we should talk with members of vulnerable groups to discover if they want or need this protection. We should also determine if harm to the subjects is likely, how bad it might be, what restrictions may be necessary, and how the least invasive restrictions can be used.

Lacking Capacity to Give Consent

The ethical basis for research policy with competent people is similar to the basis for policy about people who are not able to give consent. The main ethical concerns are strengthening their ability to choose for themselves, their fair treatment, and their well-being. There are four important policy options that give different approaches to balancing what is fair, what protects incompetent people the most, and what supports whatever ability to choose that they have.

The "surrogate," or "libertarian," solution. One policy allows the same sort of research with people who are unable to give consent as with other subjects, as long as their guardians give consent. Since guardians have the authority to choose the method of care, religion, and education of the people in their care, they should also be able to determine, according to this view, whether the people they are in charge of can participate in research.

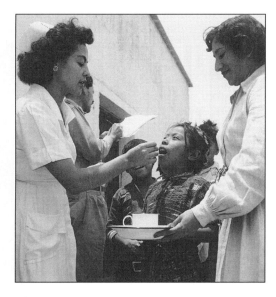

▲ Two Guatemalan nurses hand out pills to children at a school in Guatemala City during an experiment regarding the efficacy of Vitamin B-12 treatment.

The "no consent–no research," or "Nuremberg," solution. Another policy forbids using people as research subjects without their consent, no matter what. This view is given in the first international research statement, the Nuremberg Code, which states, "The voluntary consent of the human subject is absolutely essential."

The "no consent–only therapy," or "Helsinki," solution. A third policy allows people who cannot give informed consent to be enrolled as research subjects only if the studies are a part of helpful treatment and if guardians consent. This policy option notes the difference between clinical or therapeutic research and nontherapeutic biomedical research. Clinical or therapeutic research aims to benefit the person through prevention, diagnosis, or treatment of disease, such as using a new anticancer drug on a patient whose cancer cannot be cured by other treatments. Nontherapeutic research seeks general knowledge and is not intended as therapy to directly benefit the patient. Testing a new painkiller on healthy patients who do not have any pain or disease is such a case. This third policy allows incompetent people to be subjects only when the research has therapeutic goals.

The "risk–benefit," or "U.S. federal regulation," solution. A fourth approach allows research on incompetent persons when the research might directly benefit them or does not place them at unwarranted risk of harm, discomfort, or inconvenience. People who follow this view should make it clear when risk is warranted or acceptable. This policy uses the evaluation of risk to decide whether it is more valuable to promote society's good by doing a study, or to protect a person's well-being and right to make his or her own decisions.

Conclusion

Institutional review boards (IRBs) should continue to protect vulnerable subjects, but at the same time make it possible for important research to continue. Without safeguards, vulnerable subjects may be exploited. But safeguards that are too strict can have dangers of their own. They can stop researchers from learning how to improve medical care for the very groups they want to protect.

When subjects are able to give legal consent but are vulnerable to being pressured or manipulated, their consent should be monitored to make sure it is really voluntary. Most people agree that whenever possible, research subjects should be competent adults, and that when incompetent people are used as subjects, the research should be related to their health-care needs. Their guardians should give consent, and if possible, the subjects should give consent to the best of their ability. Since there are difficulties with each of the four policies on subjects who cannot give informed consent, IRBs will have to look at issues of utility, fairness, and protection without very good guidance.

Related Literature

In Nathaniel Hawthorne's story, "Rappaccini's Daughter" (1844), the main character is an apparently mad scientist who uses his daughter, Beatrice, as a guinea pig, confining her to his garden, which is full of poisonous plants, and gradually building up her immunity to the plants. That process makes her poisonous to others, and she ends up accidentally poisoning the young Giovanni who has fallen in love with her.

◆◆◆

Miklos Nyiszli, a Hungarian Jewish physician, wrote *Auschwitz: A Doctor's Eyewitness Account* (1946, trans. 1960). It describes his work as chief pathologist working under Dr. Josef Mengele in the notorious concentration camp. The memoir raises many questions about the roles and responsibilities of physicians in times of war.

◆◆◆

David Feldshuh's play, *Miss Evers' Boys* (1990), describes the infamous Public Health Service experiments at Tuskegee, Alabama, during which African-American men with syphilis were allowed to go untreated for years even after the cure (penicillin) had been found. The play portrays several of the research subjects, their Public Health nurse, and the PHS researchers, who never permitted the men to get treatment because the researchers wanted to study the long-term effects of untreated syphilis.

PATENTING ORGANISMS

A patent is a document that gives someone the right to be the only one to produce, sell, or profit from an invention or process for a certain number of years. This means that patents give inventors the right to keep other people from making, using, or selling their inventions for a limited period of time. The U.S. patent system promotes new technology through financial encouragements for new ideas and inventions. The system also promotes progress through its disclosure requirement. In exchange for exclusive rights, people who apply for a patent must describe their inventions in terms that will enable others who are skilled in the field to make or use them. When the U.S. Patent and Trademark Office (PTO) gives someone a patent, an explanation of the invention becomes available to the public. When the patent expires, the public is free to use the patented invention itself.

▶ Patent certificates for various pharmaceutical products on the wall at the Bristol Laboratories, Syracuse, New York.

Background and History

As financial interest in biological products and processes has grown, inventors have tried to patent inventions involving living materials. These inventors have met with a number of difficulties under traditional patent law. For a long time, the United States has not granted patents to products of nature and to occurrences of nature. Even if these natural products have been newly discovered, they are not considered patentable. This decision was first handed down in the 1948 case of *Funk Brothers Seed Co.* v. *Kalo Inoculant Co.* But this case and other decisions handed down before the explosion of modern commercial biotechnology made light of this hurdle to obtaining patents. In fact, the PTO granted patents on materials derived from natural sources through human intervention. The purification of naturally occurring products, such as adrenaline, was given patents so long as the patent claims did not cover the product in its natural state.

Evolving standards. Over the years, drug companies have obtained U.S. patents on such processes as using strains of bacteria to produce antibiotics. Until the 1980 decision of the U.S. Supreme Court in *Diamond* v. *Chakrabarty*, the strains of bacteria themselves were not given patents. But in this case, the Court held that a living single-celled bacterium, transformed with DNA plasmids (small circles of bacterial DNA) through human intervention to give it the capacity to break down crude oil, could be patented as a "manufacture" or "composition of matter." In arriving at this decision, the Supreme Court stated that the patent law allows patents on "anything under the sun that is made by man."

hybrid offspring of two animals or two plants of different breeds or species

With the Supreme Court's decision, the PTO quickly expanded the categories of living subject matter that it considered eligible for patent protection. In 1985, the PTO held that **hybrid** corn plants were eligible for standard patents. Two years later, it held that laboratory-altered oysters fell within the range of patentable subject matter. Soon after, the Commissioner of Patents issued a notice stating that the PTO "now considers nonnaturally occurring nonhuman multicellular living organisms, including animals, to be patentable subject matter."

Restricting animal patenting. Any claim covering a human being is not eligible for patent protection. Because the grant of a limited but exclusive property right in a human being is prohibited by the Constitution, humans are not considered patentable. But in 1988, Harvard University received the first patent on a genetically altered animal (U.S. Patent No. 4,736,866). This patent allowed Harvard to develop the so-called oncomouse. This mouse carried a human gene that lowered its resistance to cancer. This decision to extend patent protection to animals created much controversy. Numerous hearings took place in the U.S. Congress, and Congress proposed restrictive legislation on animal patenting.

While national patent laws vary somewhat, the range of biotechnology inventions that may be patented outside the United States is generally more restricted than it is under U.S. law. Patent protection is widely available for microorganisms, such as bacteria and viruses. In addition, many nations offer various forms of patent-like rights to plant breeders.

Objections to Patenting Organisms

In 1993 the U.S. Department of Commerce filed patent claims on cells taken from Solomon Islanders (people living in the South Pacific), and in 1995 from Hagahai tribesmen of Papua, New Guinea. The tribes had agreed to the patents and received royalties.

It is important to see that some of the objections to patenting living organisms—plants and animals—are really objections to the underlying technology itself, rather than to its protection under the patent laws. Such objections include fears about the hazards of altering genes (which pass on hereditary traits) in animals and human beings. Worry about the effects of this genetic engineering on public health and the environment is great. In addition, concerns that gene animal research involves cruelty to animals or reveals an arrogance on the part of human beings have been heard in Congress, in universities, and in other places. Religious organizations have warned that scientists who tamper with genes—the blueprint for life-forms—are "playing God" or violate the sacredness of life. Concerns also exist about the possibility that genetic technology might be used unethically.

On the one hand, one might question whether these sorts of objections are really the concern of the patent laws. Some believe a solution might come through the regulation of laboratory research or animal rights—or by completely preventing certain types of genetic research. On the other hand, withholding financial rewards may be an effective means of slowing the pace of research without preventing it altogether. Outright bans on research raise troubling questions about the proper role of government in research science. Denying patent protection to certain kinds of research would probably be the better route to take.

Patents' influence on research procedures. Other objections speak more directly to the effects of patenting itself. One set of concerns involves patents' impact on the conduct of research science, particularly in biomedical research. Patents could slow scientific progress by allowing patent holders to withhold important discoveries from their competitors. In addition, by encouraging researchers to be alert to the financial potential of their work, the patent system may slow scientific progress by imposing secrecy on discoveries.

The financial value of research. The patent system may also distract universities from their mission of expanding research and knowledge in favor of going after moneymaking patent rights. Of course, some of these distortions arise from the financial value of the research rather than from the availability of patent rights. In the absence of patent rights, one might expect greater interference with scientific communication as researchers seek to keep their discoveries secret. But the idea of obtaining patents may actually *increase* commercial interest in certain research. Such financial interest would probably heighten tension between science and commerce.

Another set of arguments concerns the impact of patents on various economic interests. Some critics argue that the higher cost of patented livestock or seed could bankrupt small farms.

Ethical concerns. In a broader sense, some argue that patenting life-forms promotes an unhealthy, disrespectful, or materialistic view of life. One strong argument holds that looking at a life-form as something that can be patented reduces a patented organism to nothing more than a material object. In other words, patenting life-forms would cheapen our attitudes toward life since people would treat them as goods to buy and sell.

These concerns are particularly great when the organisms to be patented have been altered to possess human traits. The creation and patenting of **animal-human hybrids** may even call into question people's ideas about the unique character of the human species. Such a hybrid, it is argued, would be a threat to human dignity. The patenting of species with human traits also triggers concerns about the effectiveness of the Patent and Trademark Office's policy to exclude human beings from being patentable subject matter. As scientists move toward creating genetically engineered humans, new concerns and questions will surely arise.

Through genetic engineering, an American pharmaceutical research company has placed a human gene suspected of causing Alzheimer's disease into a mouse. The offspring of this mouse do in fact have memory difficulties. They cannot find their way through a maze. This mouse line is the first to allow researchers to study Alzheimer's disease in an animal model. The line has been patented, and researchers spend large sums of money to procure Alzheimer's mice.

animal-human hybrid an animal that has been altered to contain a human gene

Sex and Gender

GENDER IDENTITY AND GENDER-IDENTITY DISORDERS

chromosome a rod-shaped body that contains genes of an organism; different organisms have differing numbers of chromosomes to hold all the genes

hormone a substance secreted by a cell, tissue, or organ, which circulates in the body and stimulates functions in certain other cells, tissues, or organs

Gender usually means the male or female identity of a person. An understanding of gender is necessary to understand gender-identity disorders. These disorders have to do with conflicts in being considered male or female. Knowledge about gender will also enable us to see ways in which the idea of gender is important to health care and medical practice.

Despite the temptation to view all human beings as either male or female, human sexual differences are not easily put into these simple categories. Identifying human gender must take into consideration **chromosomes**, **hormones**, anatomy, psychology, and society. Given the varied nature of such components, males are not all identical copies of one another, and neither are females. Further, the male and the female are not always especially distinct from one another. For example, some human beings are born with an extra sex chromosome. This results in XXY or XYY sex chromosomes, instead of the usual XX for females and XY for males. The sex chromosome XXY produces adults who are eunuchs (men who have no testicles). The sex chromosome XYY produces adults who appear male.

The range of human sexual traits extends to psychological qualities and social behavior. Gender identity is therefore a complex idea with biological, psychological, medical, cultural, and social components.

Getting the Terms Right

The word "transgendered" is a catchall term that describes people who live as the opposite sex, whether or not they have had sex-change surgery. This term encompasses cross-dressers, otherwise known as transvestites. Some lesbians and homosexual men describe themselves as transgendered.

The word "transsexual" is typically reserved for individuals who have had at least some sex-change surgery and take hormones to further that change.

Getting the Numbers Right

According to the International Foundation for Gender Education in Waltham, Massachusetts, about 2 percent of Americans describe themselves as transgendered. Of these, about one-eighth are transsexual, meaning that about .025 percent of Americans identify themselves in this category.

What remains to be accomplished by greater attention to gender is the larger task of identifying ways in which biology and

George W. Jorgensen, in military uniform, became Christine Jorgensen in the 1950s.

medicine can take into account people's interests regardless of gender-related differences. The idea of gender is important to an ethical analysis of identity, health, and sexual orientation.

Gender–Identity Assignment

Sometimes a newborn baby's traits do not permit an easy classification as male or female. Children may be born with irregular sexual organs and uncommon chromosome traits. Gender assignment is made by the parents and doctor to decide if a newborn is either a boy or a girl. The traits that are taken as establishing a child's "real" gender reflect biomedical knowledge. But they also reflect cultural standards about what traits are most important in deciding what makes up a boy or a girl.

Doctors have tests to determine the presence of testicles (the two sex glands in the male that produce sperm and male sex hormones), ovaries (the two glands that store and develop eggs and produce sex hormones in women), and various levels of hormones in children with uncertain sex organs. In addition, doctors have surgical methods for enlarging very small penises. Throughout history, children have been matched to one gender rather than another on the basis of physical resemblance or on the presence or absence of testicles and ovaries. Such methods could not, however, distinguish humans with true intersexed conditions, in which tissue from testicles as well as ovaries is present. Also, these methods failed to be sensitive to the complex issue of sexual identity. Chromosome and hormone tests offer another way of characterizing the gender of children. Such methods assume beforehand that human beings are divided neatly into male and female and that children of uncertain sex need to be studied to determine their place in this neat category of boy or girl.

Surgery on intersexual children involves removing and remolding genital structures, and may require the addition of parts taken from elsewhere on the body. In 90 percent of cases, surgeons choose to make the baby into a girl. These operations are usually performed shortly after birth, at the age of 6 weeks to 15 months. Several support groups oppose genital surgery on intersexual people, especially because it is done at an age when the individual cannot make an informed choice.

Decisions about gender assignment and therapy make it clear that gender can be as much a product of human choice as of biological traits. Use of the term "assignment" suggests that gender is given to a person rather than discovered. Even if the opportunity is offered to parents to assign a gender to their children after birth, medical theories suggest that a child of uncertain sex can have its gender molded through medical and psychological methods.

Gender-Identity Disorders

Gender-identity disorders are the differences between the gender one feels oneself to be and the gender that has been assigned. Most people feel they are male or female and express themselves in ways that meet with social expectations and with roles given to that gender. At a very early age, however, small numbers of individuals find that they are uncomfortable with the gender identity assigned to them because their body tells them they are a different gender. The discomfort these people feel is not just a sense of being inadequate in a particular gender role. It is discomfort with being identified as one gender rather than another.

Many contradictory theories have been given to account for being dissatisfied with one's gender. According to the American Psychiatric Association (APA), these theories suggest inherited traits, family situations, early physical illness, and early psychological events among the causes for gender dissatisfaction.

The nature of the response on the part of parents and health professionals to gender dissatisfaction is of moral or ethical interest in a number of ways. It is interesting to note, for example, that the chief disturbance given to gender-identity disorders is the conflict between the people affected and their peers and family. Children and adolescents who are unhappy with being male or female often face depression and lost opportunities as a result of their nonconforming beliefs and behaviors. For this reason, gender-identity treatment is considered morally right because measures to protect a child from conflict and depression are the responsibility of a parent. However, a person would still need to consider the extent to which children and adolescents might be treated because they live in a society that believes in separate gender roles, even though such roles are not true to the diversity of human beings.

Transsexualism

The term "transsexual" was first used in the 1940s to describe adults who identified themselves as male or female in contradiction to the sex of their body and the behavior expected of that sex. Historically, transsexuals were thought to suffer from psychological disorders. Some health-care professionals still hold this view. Transsexual therapy describes a range of therapies that include

▲ Adolescent dressed in both male and female outfits, illustrating gender identity confusion.

surgery, hormone treatment, and psychological counseling. The goal of these therapies is to conform secondary sex characteristics (such as the growth of a beard for a man) and behavior to those of the desired gender. This therapy has important implications for a person's legal and social standing, including the right to marry.

Several ethical questions surround the practice of transsexual therapy. First, there is the request to have one's sex changed. Such requests suggest that one's gender cannot be reduced to sex organs or chromosomes. To move away from seeing transsexualism as a psychological problem is to recognize that psychological identity and biological sex identity are separate and that some aspects of gender are a matter of choice. Even if it is acceptable that gender is not just any single trait, it may not automatically follow that transsexual therapy, surgery, and lifelong hormone treatments are ethically right.

Attempting to find ways to conform body and behavior to a different gender might promote the general good of all people insofar as it leads to important biomedical advances in addition to benefits for particular individuals. It is unclear that transsexual therapy affects others in ways that are so harmful as to justify taking away the legal or medical means that would enable a person to change his or her sex.

Sexual-Orientation Therapy

Since the coining of the term "homosexuality" in 1869, certain techniques have been used with men and women to redirect their sexual orientation from homosexuality (feeling sexual attraction for a person of the same sex) to heterosexuality (feeling sexual attraction for a person of the opposite sex). Much of this therapy has been based on a strict identification of male and female with desire for the opposite sex. According to this therapy, maleness means attraction to women. Femaleness means attraction to men. Departures from this gender identification have been treated in a variety of ways. These ways include both moral and legal disapproval and medical ideas about homosexuality as a disease. Treatment methods used to make a person heterosexual include various kinds of behavior therapy, drug and hormone treatment, surgery, and a wide range of psychological therapies.

Homosexuality no longer considered a mental illness. In 1973, the APA abandoned the view that homosexuality was a mental illness. The association had held that homosexuality was a mental disease since its first manual in 1952. The APA nevertheless maintains the category "sexual-orientation distress" for individuals who are disturbed by their homosexuality and wish to become heterosexual. Despite the fact that the APA no longer views homosexuality as a mental illness, some therapists still argue that homosexuality is a disease. Other therapists wish to treat homosexuality, but they do not

Holly Dover, professor of sociology at the University of Victoria in Canada, is a lesbian who considers herself transgendered (she lives as a male). In her book *FTM: Female-to-Male Transsexuals in Society* (1997), she tells the stories of forty-five transsexuals, often in their own words.

Jay Prosser, a transsexual professor at the University of Leicester in England, wrote *Second Skin: Body Narratives of Transsexuals* (1998).

see also

Sex and Gender: Homosexuality

▶ A 1919 photo montage by the German artist Hannah Hoch (1889-1978) satirizing contemporary images of men and women.

identify it as a disorder. Some counselors who practice from a religious point of view treat homosexuality as recovery from sin rather than from disease. As part of the goal of these religious reorientation programs, these therapists use Bible study, prayer, and group socializing to restore heterosexuality. Sexual orientation, then, is an important opportunity to consider what such therapy suggests about a society's understanding of gender, difference, suffering, and medical or social injustices.

Critics of sexual-orientation therapy. Some people have argued that sexual-orientation therapy is immoral because it contributes to social prejudice against homosexual men and women. Many gay men and lesbians see efforts designed to bring about a change in sexual orientation as a continuing disregard for their lives or for the worth of homosexuality in general. Even if sexual-orientation therapy does not contribute to social injustice directly, the pursuit of sexual-orientation therapy may be the result of social injustice rather than an injustice itself. Such social injustice stems from the view that homosexuality is undesirable and inferior to heterosexuality.

Psychoanalysis and sexual orientation. Reports of success in changing homosexuals to heterosexuals usually come from psychoanalysts. Psychoanalysts are doctors who believe, as Freud did, in having patients talk freely as a way of analyzing conflicts. These claims also come most typically from therapists who work on changing behavior. These reports of success have been criticized for the following reasons: the small number of sample patients; success standards that measure behavior change but not psychological or emotional readjustment; inadequate long-term assessment; and the uncontrolled way in which the studies were handled.

Ethics of sexual-orientation therapy. While there are reports of success in achieving heterosexual behavior and even heterosexual marriage, it is still unclear whether the therapies themselves have been responsible for these changes or whether these changes will last for a person's lifetime. Because of these problems, it is unclear whether there is any really effective therapy that is useful on randomly selected persons seeking sexual reorientation. If this is so, patients should be told that such therapy is experimental and unproved. In the past, some homosexuals were treated without their consent. In view of the respect we have for an individual's right to make his or her own decisions and the modern view that homosexuality is neither a moral nor a medical ill, involuntary treatment of homosexuality can only be seen as unethical and wrong.

Conclusions

Gender identity has become a focus for moral and political analysis both in medicine and in the culture at large. This analysis usually centers on the ways in which the biology and cultural construction of gender are intertwined. This analysis asks whether and with what consequences male gender stands as the model for all human beings. Moral and political analysis also questions whether polarized notions of gender in biology and society are adequate to express the range of human gender diversity. Also, it asks whether the standard of heterosexuality is valid for judging the nature and destiny of all people. This analysis makes clear that treating gender, social roles, and sexual orientation as undivided in nature often has led to injustices against women, intersexed beings, and gay men and lesbians, who thus suffer oppression and disadvantages on the basis of gender. An ethical analysis of gender differences will offer biology and medicine the opportunity to serve human beings as the persons they are rather than as the persons society, medicine, or politics would have them be.

The University of Iowa, Rutgers University, and others are establishing policies to protect faculty and students from discrimination based on gender identity. The Transgender Task Force, a group of students at Harvard University, convinced the student undergraduate council to include "gender identity or expression" to the list of protected categories in antidiscrimination policies. In 1997, Harvard decided to allow a transgendered student who dresses like a man but is biologically a woman to live on an all-male floor of a campus dormitory.

In 1969, the Kinks, a British rock group, had a pop hit about a transgendered transvestite named "Lola." The popularity of the song was a bit surprising, with lines like "Boys will be girls and girls will be boys."

Related Literature

Pamela White Hadas's poem "To Make a Dragon Move: From the Diary of an Anorexic" (1994) is narrated in the voice of an anorexic teenager. The young woman claims she is *king* of her body and describes with horror the look of her mother's "tapioca" thighs. Her anxieties about fat and food, as well as sex and growing up, all compound to make her want to disappear, to become nothing but bones, to starve herself to death rather than allow her body to mature into womanhood.

HOMOSEXUALITY

psychologist a nonmedical professional who treats patients suffering from emotional problems

Religious traditions place both negative and positive values on homosexuality. In some traditions, homosexuality is strictly forbidden, while in others it is associated with supernatural powers. In warrior cults in ancient Greece, homosexuality was a part of male-bonding rites.

In order to understand homosexuality, it is important to know the definitions of certain terms. A homosexual is someone who is sexually attracted to a person of the same sex. A bisexual is a person who is sexually attracted to both men and women. A heterosexual is someone who is physically attracted to a person of the opposite sex. Labels such as homosexual, bisexual, and heterosexual have meaning within a social, political, and cultural context. As social, political, and cultural views change, our understanding of the meaning of such labels changes, too. This point is important for **psychologists** and other health professionals to keep in mind when assessing a person's sexual orientation and in finding the words to describe it.

Throughout history, doctors and others have based a person's sexual orientation on the genital organs of the individual and on the sex organs of the individual to whom the person was sexually attracted. Many health professionals have also believed that a person's sexual preference is fixed and unchangeable. Recent research, however, has challenged these ideas. We now recognize that a person's sex—male or female—is only one of the many components of the attraction two individuals feel in relation to each other. So we must understand that sexual orientation is many-sided. It takes into account a person's sexual identity (male or female physical characteristics, male or female gender identification, traditional masculine or feminine roles), as well as behavior, fantasies, and emotional attachments.

Theories on the Cause of Homosexuality

Throughout the ages, doctors and others have expressed many theories about the development of a person's sexual orientation. But no one theory exists today.

psychoanalyst a person who believes, as Sigmund Freud did, in having a patient talk freely as a way of analyzing problems

Among the Navajo (Southwest), Cheyenne (northern Plains), and Mojave (southern California) Native Americans, "berdaches" were men dressed as women. They engaged in homosexual activities, usually with married men. Berdaches were responsible for many religious rituals, including cutting the ritual lodgepole and officiating at scalp dances.

Childhood trauma. Traditional psychological theories have suggested that homosexuality is caused by some childhood trauma or emotional shock that has had a lasting effect on a person's mind. This effect has stopped psychological and sexual development and causes the person not to be attracted to the opposite sex. Many psychologists still hold this view. But this traditional view of the cause of homosexuality has been challenged by many scientists, including homosexual ones. In his book *Being Homosexual: Gay Men and Their Development* (1989), Richard Isay, a well-known **psychoanalyst**, believes that homosexuality is inborn and influenced by biological factors.

Learned behavior. Another traditional view has been held by specialists in learning and behavior. These people have traditionally seen homosexuality as an illness or as a poorly adapted learned behavior. In fact, behavior specialists believe that homosexuality is learned through conditioning.

Variation of sexuality. In contrast, in his book *Gay, Straight, and In-Between* (1988), John Money, who has studied the causes of sexual attractions for more than forty years, writes that homosexuality is a normal variation of sexuality. Money states that influences taking place before birth interact with environmental events at critical periods and cause an individual to become homosexual. The exact way in which these influences work is unknown.

Biological factors. In a major study of the development of sexual orientation, *Sexual Preference: Its Development in Men and Women* (1981), Alan Bell, Martin Weinberg, and Sue Hammersmith were unable to confirm any of the psychoanalytic or behavioral theories among the large sample of people they studied. While these three authors were not able to find an environmental influence, they suggested that research should focus on biological or inborn factors to find the cause of heterosexuality and homosexuality. As a result, in the 1980s and 1990s, a few studies focused on biological factors (hormones, genes, etc.) as a cause of adult sexual orientation. Some studies were even able to identify a potential genetic component. This would mean that homosexual behavior is determined by genes and, therefore, inherited. But without solid repeated studies, the influence of biological factors is still very unclear and much debated.

Beyond the Illness Model

Since Alfred Kinsey's pioneering work in the 1940s and early 1950s, the linking of homosexuality with illness has been gradually proved to be wrong by scientific research. This scientific evidence led to the American Psychiatric Association's 1973 decision to declassify homosexuality as an illness. Some psychiatrists still argue about whether this decision was based on science or on politics. The concept of some homosexuality as an illness was nevertheless kept by the use of the term "ego-dystonic homosexuality." This term describes a disorder in which individuals have recurring, troubling homosexual desires and want to become heterosexual.

In spite of these changes in scientific opinion and the removal of homosexuality as a mental disorder, there is no complete agreement on the matter. Some professionals continue to view homosexuality as the result of an abnormal process of development that is a form of disease.

The Development of Sexual Identity and Sexual Identification

The process of developing sexual identity is probably a complex mix of biological, psychological, and social influences about which we have little understanding. Therefore, scientists have developed various models that suggest that people acquire a homosexual identity in stages. These stages illustrate how individuals who are for the most part homosexual construct their identity in a positive manner, despite living in a society that views homosexuality as a disgrace.

After a period of confusion and a sense of being "different," a person adopts a homosexual label and begins a process of exploring his or her sexual identity. The person then begins to explore his or her sexual relationships and what it means to be intimate with another person. Finally, the person accepts himself or herself for who he or she is. Taking the label "gay" (often used for homosexual men), "lesbian" (used for homosexual women), or "bi" (used for bisexual men and women) often helps the person develop more positive attitudes about being homosexual. Although the process of forming one's sexual identity is similar for men and women, men are likely to experience more sexual relationships during their exploring stage than women. Women are more likely to define who they are based upon awareness of their emotional attachments.

The psychological problems that homosexual men and women face result from social stigma or disgrace rather than from sexual orientation. The effects of being stigmatized are well understood. As long as society stigmatizes homosexuality, homosexual men and women are at risk of poor psychological and social adjustment and development. They are also at risk of family alienation, inadequate relationships, alcohol and drug abuse, depression, suicide, and anxiety over AIDS and other sexually transmitted diseases. In contrast, a positive reaction to sexual diversity on the part of society is linked to psychological and physical health.

In opposition to same-sex marriage, U.S. Congressman Robert Dannemeyer, a Republican from California, stated: "God created Adam and Eve, not Adam and Steve."
(Quoted in *USA Today*, May 1, 1990.)

In *Heeney v. Erhard* (1995), the Minnesota district court ruled that the Minnesota Human Rights Act does not require doctors to provide artificial insemination to a lesbian couple.

Ethical Issues in Psychological or Psychiatric Treatment

The ethical issues in the treatment of gay men and lesbians are for the most part no different from the issues that a psychotherapist or counselor meets in treating people seeking other types of help. The issue of "converting" to heterosexuality remains an ethical issue for those people who view same-sex attraction as a sickness. In fact, there is no evidence that adults are able to maintain long-term change in sexual orientation through therapy. Changing behavior has had limited success and usually does not last. Still, some individuals who have homosexual attractions want to "convert" to heterosexuality. Parents who are concerned that their children might be gay or lesbian sometimes will encourage them to seek counseling to ensure that they will be attracted to the opposite sex. Some therapists continue to accept patients into therapy with the goal of changing their sexual orientation. But other experts have rejected these goals of treatment as unscientific, unethical, and psychologically scarring. Nevertheless, some psychoanalysts still consider it possible to change a person's sexual orientation. In fact, they see this goal as being ethical, right, and justified as long as the individual wants to change.

Of course, the attitudes of health professionals toward homosexuality greatly influence their ability to help their patients. Health professionals can create an atmosphere for their patients in which sexuality can be discussed openly. By raising the issue of sexual orientation, the therapist can address and resolve the patient's problems and concerns. Like other people, health professionals are affected by society's attitudes and values concerning homosexuality. This means that while they may hold intellectually positive attitudes, their emotional responses toward homosexuality may be hindering them. Even a health professional may not be expressing full acceptance of a patient's homosexuality. He or she may not really be encouraging the patient to explore a positive, integrated homosexual identity.

In Hawaii, a local court ruled that homosexual couples can be legally married. The case has been appealed. No marriage licenses may be issued until the Hawaii Supreme Court settles the case.

Related Literature

Audre Lorde constructed *The Cancer Journals* (1980) out of her diaries, poetry, and prose. Lorde underwent a mastectomy for breast cancer. As an African-American lesbian and active feminist, Lorde resisted the pressures to wear a false breast or go through reconstructive surgery to look more "normal" to the male world. Instead, she urged women to avoid all such pressures and to wear their scars proudly, like the women warriors who cut off a breast to be more free to use bow and arrow.

SEXISM

Many feminist writers argue that the oldest forms of religion involved worship of the earth mother goddesses. However, the goddesses were replaced by male deities, especially after the rise of Jewish patriarchal monotheism (belief in one God). In modern times, some women have returned to the worship of the earth goddess, rejecting the male God of Western tradition.

Wicca is an old English term for the different traditions of Neo-Pagan Witchcraft, a nature religion that worships the goddess of life and fertility. Followers of Wicca believe that the natural cycles of a woman's body place woman closer to nature and to God (male or female).

Sexism is the belief that one sex is superior to the other. It also includes the many consequences of this belief. The term "sexism" almost always refers to men's claim that they are superior to women. The belief that women are superior to men is known as female chauvinism. But women rarely express this belief, and female chauvinism does not affect men's economic or social status in the way that male sexism affects women's status. Women who are actively involved in women's rights—feminists—coined the term "sexism" in the mid-1960s.

Sexism as Natural

Opponents of feminism view sexism as an idea put forth by misguided feminists. According to these opponents, discrimination by sex is a result of the natural differences between men and women, especially in reproduction and in physical strength. In other words, women can have babies, and men cannot; men's bodies are physically stronger than women's. Men and women are naturally different. Those who oppose women's rights do not see the sex-based differences that exist in societies as discrimination. Instead, they view them as necessary for the natural functioning of human society. These opponents of feminism see "sexism" as a false concept—an attempt by women to change the natural order of things. Those who view sex differences as natural can agree with feminists that women and men have different roles. They can even agree that women have been excluded from certain areas in life. But they believe that women have not been oppressed, or kept down. Such thinkers argue that to try to fix such distinctions between men and women will cause more harm than good.

Classic Sexism

The most common explanation of sexism is that discrimination between men and women leads to women's oppression. People who claim this also believe that when the biological differences between male and female become the cultural differences between men and women, such differences turn into a cultural category that keeps women down. Feminists point to the existence of prejudices or practices that restrict women's opportunities and their lives; they conclude that discriminations affecting women are wrong.

In Western societies, feminist scholars can point to many instances of sexism directed against women, such as those that exist in the story of creation in the Old Testament. God created Adam first and then created Eve from Adam's rib, suggesting an inferior status. The Greek philosopher Aristotle, in the fourth century B.C.E., doubted that women were capable of reason. Roman law made the father the head of the household.

▲ British policemen arrest a suffragette (defender of woman's rights) in 1913.

Contemporary Sexism

Modern-day institutions continue to keep women from certain spheres of life. Women hold fewer public offices. They cannot enter the priesthood in some religions. Women are kept out of some military schools or experience sexual harassment in the military. Women are less educated than men and hold fewer university jobs. Few women become officers in corporations, and for the most part, companies do not consider them as important as their male colleagues. Fewer women than men are economically independent, and women remain largely responsible for the family and chores around the house. Even in societies that boast of greater equality for women, the exclusion of women from certain aspects of life still exists. Violence against women is an ongoing problem. Women have also been excluded throughout most of Western history from important roles in cultural production, including art, literature, and music, or have been treated as inferiors in cultural work.

Development and sexism. Throughout the world in the nineteenth and twentieth centuries, feminists have struggled against sexism. But many cultures still practice ways of controlling women and their sexuality. Changing patterns of employment and consumerism in a global economy seem to have particularly harmful effects on women.

Within this account of classic sexism, different views exist about the origin and nature of the oppression in sexism. Some liberal feminists see the oppression of women as a sum total of individual acts of discrimination. They view sexism as a result of continued prejudice. In other words, the continuation of sexism is the result of societies' not being able to get rid of some of their outdated traditional beliefs about the differences between men and women. However, liberal feminists who favor reform believe that, although sexism has deep historical roots, it can be eliminated through careful remolding of social institutions.

Childrearing roots of sexism. Another strand of liberal feminism claims that sexism comes from Western practices of childrearing. Children's psychological development, including their sense of gender (being male or female) identity, depends on how they interact with caregivers in the first years of life. Because women do most of the caregiving to infants, boys and girls have a different psychological relationship to these early caregivers. Boys learn to separate from others while girls learn to bond and remain connected. These gender differences continue to tell much about the

Many feminist theologians believe that until the Jewish, Christian, and Islamic images of God change from male to female, or at least to some image that is neither male nor female, the domination of men over women will never be completely eliminated.

▶ A billboard with a message against sexism in alcohol advertising.

differing psychology of men and women. According to this theory, eliminating the male–female division of parenting in Western society would eliminate sexism.

Institutional roots of sexism. Other feminists view the discrimination against women and the resulting oppression of them as deeply rooted in social institutions. They believe that merely changing people's prejudiced attitudes and practices will not get rid of sexism. The ideas of the German philosopher and political economist Karl Marx have influenced some of these feminists. Marx, who lived in the nineteenth century, developed a system of ideas that has served as the basis of modern socialism. According to Marx, private ownership should be done away with. Society, not private individuals, should own and operate the means of producing and distributing goods, with all members of society sharing in the work and the products.

Marxist feminism. Feminists who believe in Marx's ideas see sexism as being linked to the idea of owning private property. These feminists feel that as long as it is necessary to ensure the proper inheritance of private property, men must regulate women's sexual activities so as to produce a legal heir to the property. Marx's associate, Friedrich Engels, argued that since the exploitation of women began with property, women's oppression would end with the Communist revolution. As a result of that revolution in Russia in the early twentieth century, private property was done away with and put in the hands of society. A number of socialists have challenged Engels's argument as being too simple, and the treatment of women under so-called socialist regimes has not been encouraging.

Ethical issues. This question about the nature of classic sexism remains: Is it a matter of prejudice expressed by individuals' attitudes or beliefs, or is classic sexism a form of discrimination that exists in social, political, and economic institutions?

In the first century of Christianity, with its emphasis on equality for all believers, women were able to hold leadership roles in the church. However, due to the surrounding patriarchal cultures, women gradually were eliminated from any role with authority. Even those who did have such a role were forgotten by history.

In *The Feminine Mystique* (1963), Betty Friedan defined women's liberation. According to her, the freedom of women had been limited by men's demands of them to be sex objects and mothers and to live their lives through their husbands and children. Women were therefore unable to realize their own potential. Friedan did not perceive men as the enemy, but instead felt that they, too, would be liberated as soon as they allowed women to be full people.

Male teenagers are frequently sexist. Among certain groups it is considered a sport to have sex with as many young women as possible. For these young men, sex is not linked to love and intimacy and commitment but to conquest and winning, without any regard for the women.

This split within the understanding of classic sexism matters for the study of ethics on two levels. First, the tasks involved in solving a problem based on attitude or behavior are different from those involved in changing the very nature of social, political, and economic institutions. Second, the responsibility for sexism is quite different depending on whether it is simply a matter of prejudice or a more deeply rooted social problem. If sexism is simply a matter of prejudice, then those who believe they no longer have prejudiced attitudes against women are no longer sexist. On the other hand, if sexism is part of the social structure, then simply changing one's attitudes does not offer a solution to the problem. In the last quarter of the twentieth century in the United States, some attitudes toward women have become more favorable. But no parallel change in women's economic, political, and social standing has occurred. This fact suggests that sexism is institutionalized.

Androcentrism

Androcentrism is the belief that men's experiences are universal and that women's experiences are similar to men's. According to androcentrism, sexism does not begin from a conscious act of discriminating between men and women. Instead, it starts with the view that men's lives and experiences, their bodies, behaviors, activities, and beliefs should serve as the normal starting point when thinking about human beings. Those who argue that androcentrism is really sexism believe that by defining the male experience as normal, women's experiences, lives, bodies, and so forth are viewed as lacking, as differing from the normal.

An example of androcentrism is the medical practice of considering male bodies as normal. Doctors and medical researchers often exclude women from clinical trials because female hormones complicate medical observations. Women may then receive less medical attention. As a result, doctors actually know less about how disorders such as heart disease affect women.

Conclusion

These different accounts of the nature of sexism have a deep effect on how bioethics might address the problem of sexism. If one takes a classic, liberal view of sexism, then removing one's own prejudice is enough. If sexism is deeply rooted in social, political, and economic institutions, then one must be aware of its effect throughout those institutions. Economic oppression may affect access to medical care. If sexism is androcentrism, then doctors, researchers, and other health-care workers must constantly be on their guard to make certain that practices and experiments do not view the male as normal and the female as straying from the normal. Those involved in bioethics must guard against sexism even though they do not openly discriminate against women.

 This image, created by Barbara Kruger, sends a message about a woman's body being the subject of political and social conflict.

Related Literature

In Kate Chopin's novel *The Awakening* (1899), a young woman evolves from being a dutiful wife and mother into a more independent character who wants to paint, to swim, and to act as she wishes rather than to fit into the "proper" female roles. Defying her husband, she sets up her own household without the children and indulges in a romantic affair. But when she feels like a total misfit, she drowns herself in the sea.

◆◆◆

In Alice Walker's novel *The Color Purple* (1982), both men and women are conditioned to think they have to fulfill certain roles assigned to their sex. Celie believes she must submit to her husband, take care of the children, cook, clean, sew, work in the garden. It takes the experience of Shug to help Celie break out of that trap. The men, too, are hurt by assuming they have to fulfill certain sex roles.

SEX THERAPY AND SEX RESEARCH

This entry consists of two articles explaining various aspects of this topic:

    Scientific and Clinical Perspectives
    Ethical Issues

SCIENTIFIC AND CLINICAL PERSPECTIVES Sex has many functions. Sex allows people to have children. Sex makes people feel that they are worthy and desirable, and it can be a strong bond between them. Many people see sex as more than just a natural physical act. They believe sex is a special gift that can show love and affection. Some people see it as a way of controlling others. And some believe sex is evil and should be punished.

Each culture has its own beliefs and attitudes about what is normal and right and what behavior is best. These beliefs affect people's attitudes toward sex and sexual behavior. Beliefs and attitudes about sex have been very different in different times and places. Because of this, the fields of sex therapy and sex research have been affected by questions about what is right, healthy, and ethical.

Sex Counseling and Therapy

Sex counseling mainly involves educating people about sex. Usually people come into counseling with a particular sexual concern or

The ancient Hippocratic oath clearly forbids any sexual relationship between doctor and patient: "Into whatever houses I may enter, I will come for the benefit of the sick, remaining clear of all voluntary injustice and of other mischief and of sexual deeds upon the bodies of females and males, be they free or slave."

The American Psychiatric Association's *Annotations Especially Applicable to Psychiatry of the Principles of Medical Ethics* reads: "Sexual activity with a patient is unethical." This is the only absolute prohibition in the document. State medical associations have suspended the medical licenses of a number of psychiatrists who have been sexually intimate with their patients.

problem. The counseling focuses on solving this problem and usually requires the person to visit the counselor only a few times.

Sex therapy is more complex. Some sexual problems take longer to understand and manage. Sex therapists look at the problem and how it fits into the patterns of the person's personality and relationships with others. Although education is part of the therapy, other treatment methods are also used.

The goal of sex therapy is to solve sexual problems. Usually these problems involve such physical aspects of sexual response as desire, excitement, or orgasm. But sex therapy can also involve treating people who express themselves sexually in ways that society disapproves of, such as people who expose themselves to others or who want to force others to have sex.

Sex counseling. Some clients need only information. Sometimes people have incorrect information about sex that simply needs to be corrected. Some need to be reassured that they are normal. Others have been hurt by society's unrealistic attitudes about what kind of body is beautiful and what kind of behavior is normal. Our society praises thinness so much that some overweight people may feel sexually unattractive and may need to talk to a therapist to get a more realistic view of the situation.

Some people find it difficult to talk about sex with their partner. They may not know what words to use, they may be embarrassed, or they may be afraid of hurting the partner. These couples can come in together, and the therapist can help them talk about what they want, need, and expect from sex.

Many people are looking for encouragement and advice to help them sort out the stressful problems of their lives, which make them so busy that they do not have the time or energy for sex. Some may have experienced changes in their health, with conditions like arthritis or a recent heart attack, that affect their sex life. Others want to learn new ways to experience sex with their partner, to enrich their sexual life.

The therapist may educate the client and try to change attitudes and behaviors that make the client anxious about sex. The most common model used for this education and change is called the Sexual Attitude Restructuring Process. This is a workshop developed by the National Sex Forum. People attending the workshop look at sexually explicit media and then have a group discussion about their reactions to it.

Approaches to Sex Therapy

Most of the methods used in sex therapy come from three approaches: the therapeutic approach, the behavioral approach, and the medical approach. Some sex therapists use a combination of all three.

Therapeutic approaches. These therapies are based on the idea that sexual problems are caused by psychological problems. People who follow this approach usually believe that sexual problems come from deep conflicts that the client experienced when he or she was a child. In therapeutic treatment, the therapist and client work to find out and understand the patient's conflicts. This kind of treatment focuses on understanding these conflicts and making the whole personality healthier, rather than just fixing the sexual problem.

But this kind of therapy generally lasts a long time and is expensive. Also, it is difficult to measure how well it actually helps people with their sexual problems. Other approaches may put less emphasis on childhood problems, and they may spend more time on the sexual problems that the client has as an adult such as difficulty communicating with the partner, hurtful relationships, and negative attitudes and incorrect beliefs about sex.

Behavioral approaches. Here, the basic idea is that behavior, including sexual response, is learned. It can also be unlearned, and replaced by behavior that causes fewer problems. Behavioral therapists believe that most sexual problems are caused by anxiety. They do not spend a lot of time talking about the client's childhood, personality, or relationships. They deal with sexual problems by training the client to overcome anxiety or to become less sensitive to things that cause anxiety. Most of these techniques are based on gradually exposing clients to things that make them anxious, until they no longer become anxious. Other methods involve changing a client's sexual fantasies, or training clients to associate their undesirable sexual behavior with something unpleasant, so that they no longer want to do undesirable things.

cognitive involving the process of knowing or of acquiring knowledge; pertaining to the state of awareness, perception, and reasoning

Other approaches combine elements of both behavioral and **cognitive** therapy. Cognitive-behavioral therapy is helpful in treating many disorders, and emphasizes changing people's thoughts about sex.

Medical approaches. These therapies focus on the fact that some sexual problems are caused by physical problems affecting the nervous system, circulation, or metabolism. Sexual problems can also be caused by diseases that affect the entire body, such as heart disease or arthritis. Illegal drugs, prescription drugs, or over-the-counter drugs can all cause sexual problems.

Scientific studies about how physical factors affect the sexual system have begun only recently. Most of them have been done on men, so not as much is known about women. The greater emphasis on men in research may have resulted because most of the researchers have been men. Some say that because women are better at hiding their sexual problems or because they are reluctant about discussing sexuality, it is harder to do research with them. Others of them believe women are less concerned if they lose sexual function because, for many of them expressing and receiving affection is more important than purely physical sexual functioning.

In the 1960s and 1970s, William H. Masters and Virginia E. Johnson conducted research into human sexual behavior.

The most famous American experts on sex therapy, William Masters and Virginia Johnson, were very concerned about the moral character of sex therapists. In strong terms, they argued that therapists who have sex with patients should be charged criminally with rape, no matter who initiated the seduction.

A basic attitude of caring, usually evident in the little things (tone of voice, compassion, attentive rather than distracted listening, responsivity to needs great and small), rather than sexual eros, is imperative in therapeutic relationships.

Sex Therapy

The methods that a sex therapist uses can be very different, depending on the client's problem and the therapist's beliefs about what therapy works best.

Professional practice. The classic approach was developed in the 1960s by William H. Masters and Virginia E. Johnson for use with male-female couples. This involves a two-week program of daily therapy while the couple is away from their usual work and family responsibilities. Two therapists, one male and one female, work with the couple. This is so that each partner has someone of their own gender "on their side." The couple is treated together, as a unit.

This approach is not practical for many people, because it is difficult for them to get away from work and family for two weeks or because their lives do not fit the pattern of two people involved in a male-female relationship. For such people some therapists may change their approach—by using only one therapist, treating people who are homosexual or bisexual, treating individuals as well as couples, and using group therapy. Some therapists may include a more traditional approach and talk about the client's past conflicts, especially if behavioral approaches are not working.

Most treatment programs have some elements in common. The client talks with the therapist to figure out exactly what the problem is and how it might be treated. The client may give a detailed description of his or her sexual past, including attitudes, experiences, and feelings. The client may take a test on sexual knowledge, attitudes, and beliefs. The client might also be examined by a doctor to determine if there is a physical cause for the problem.

The therapist may educate the client about sex and may give the client "homework"—behaviors to practice at home. Then at the next meeting, therapist and client discuss what the client has learned. If the client is a couple, the therapist helps them talk with each other about the problem.

Advances in sex therapy. The knowledge and basic theories in the field of sex therapy have gone through great changes. In the 1960s, Masters and Johnson did pioneering research on "normal," healthy people that helped therapists and others learn just what is normal and how physical factors shape sexual response. Before the work of Masters and Johnson, most therapists believed that sexual problems were caused almost entirely by psychological problems and anxiety. After the results of Masters and Johnson's research became public, therapists in the 1970s began to see that many sexual problems were caused by such physical factors as illness and drugs. Because of this, clients often must be examined by a doctor before they begin therapy.

As a result of research and therapists' experiences with clients, two new advances in thought and knowledge have changed the field of sex therapy. The first advance was the understanding that the sexual response does not just happen all at once but is a cycle of responses. Masters and Johnson called the parts of this cycle excitement, plateau, orgasm, and resolution. The second advance was the understanding that many different psychological and physical factors affect sexual problems. There are many different kinds of sexual problems, and each one may be treated with a different, specific plan. These two advances led therapists to come up with specific treatments for specific problems, making treatment better and more likely to work.

Whether or not a treatment will work depends mostly on two factors—the kind of problem and the right treatment for it. It is easier to treat someone who has difficulty having an orgasm than someone who has no sexual desire at all. If someone has a physical problem, sex therapy alone will not help, and medical treatment may also be necessary. If someone has been sexually abused during childhood, it is unlikely that behavioral treatment alone will help. The client may also need more traditional therapy to talk about feelings and fears about sex.

In the past, some people who were homosexual were forced to go to sex therapy or chose to go because they were uncomfortable with their homosexuality. But therapy to change sexual orientation has become ethically questionable, since homosexuality in itself is no longer seen as a sickness or treated as one. Most therapists now simply try to help homosexual clients deal with the stress of living in a disapproving society.

Qualifications for professional sex therapists. Sex therapy is a new profession. Like any new profession, it has to answer questions. Who should be allowed to be a sex therapist? What knowledge, skills, and attitudes should they have? Who should certify that they are skilled enough to practice? Who oversees the profession to be sure bad therapists are not allowed to practice? These questions have been discussed since sex therapy became a profession in the 1970s, but they are still not answered. It is clear that some sex therapists may be unqualified people who want to make money, or who have sexual problems of their own.

Sex therapy includes a wide variety of treatments, including hypnosis, **biofeedback**, and surgery. Many different kinds of people, with different skills, may call themselves sex therapists. Sometimes health professionals, who may be licensed as doctors, nurses, or physical therapists, may try to do sex therapy. But these people may not be properly trained to understand and treat sexual disorders.

Since the 1960s, doctors, nurses, **psychologists**, and social workers have all received more training about sexuality and sex therapy. But it still has not been decided who can practice sex therapy, how much training they should have, and what the training should include. As the profession matures, these questions may be answered.

see also

Sex and Gender: Homosexuality

biofeedback becoming aware of the body's responses in order to alter or control them

psychologist a nonmedical professional who treats patients suffering from emotional problems

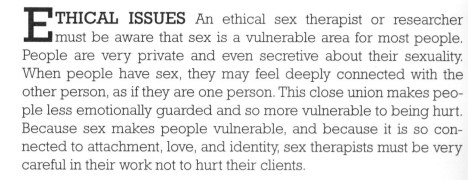

ETHICAL ISSUES An ethical sex therapist or researcher must be aware that sex is a vulnerable area for most people. People are very private and even secretive about their sexuality. When people have sex, they may feel deeply connected with the other person, as if they are one person. This close union makes people less emotionally guarded and so more vulnerable to being hurt. Because sex makes people vulnerable, and because it is so connected to attachment, love, and identity, sex therapists must be very careful in their work not to hurt their clients.

A couple with their therapist

Social Attitudes About Sex

Every culture has its own ideas about what is right and "normal" in sex. Usually these ideas are so taken for granted that no one questions them, and it is hard for people to realize they are ideas, not facts. Sex therapy developed in the mostly white, middle-class, North American scientific culture. What unquestioned values and ideas from this culture are a part of sex therapy and research? Can treatments designed for white, married couples be used with gay couples or a bisexual person? Can they be used with an African American or Asian? Do members of minority groups have problems with therapy because it is based on a culture and standards that differ from their own?

A culture's expectations and beliefs about sex and what is normal can affect people's definitions of which sexual behavior is normal and right. Sex therapists also are affected by the culture they grow up in and may follow its beliefs when they say some behavior is abnormal. For most of the twentieth century, many therapists believed that homosexuality was a mental disorder, even though there was no evidence that it was.

Social Impact of Attitudes on Individuals

Many writers on sexuality discuss how a society's attitudes about sexuality can have negative effects on individuals and groups. Beliefs about female sexuality have had a great influence on the way women have been treated throughout history. Without any real information, many societies, including their scientists, have believed very different things about women and women's sexuality. People have at times believed that women have endless sexual desire and at other times that women have no sexual desire. Some people have believed that a woman can get pregnant only if she has an orgasm during sex. Others have claimed that one kind of orgasm means that a woman is more mature, while another kind means that she is less mature. A mature woman was supposed to enjoy an orgasm only from intercourse with a male partner. Although scientists as well as ordinary people believed these things, there was never any information to show that they were true.

The Therapists' Values

Who are the therapists? What attitudes, values, and longings do they have? Since they are a part of therapy, their own attitudes and beliefs are important. They may specialize in sex therapy, or they may be doctors, nurses, social workers, or other health-care professionals who help patients with sexual problems. Whatever their training, they may have strong emotional reactions to patients and their problems. These reactions can affect their ethical decisions about what is good therapy.

Positive attitude and tolerance. If therapists believe that marriages must be kept together at any cost, that financial, emotional, or physical abuse is wrong, and that partners in a relationship should be complete equals, these beliefs will strongly affect the therapy. Therapists should have a positive attitude toward sexuality and should see it as a creative, life-enhancing force. Therapists should not have sexual conflicts of their own and should be tolerant of others' differences.

Because no one is perfect, therapists must know if they have any personal values or preferences that might get in the way of their helping the client. For example, a therapist who is uncomfortable about homosexuality should know this and should not treat gays or lesbians.

Providing care. A health-care worker's first duty to the patient is to provide care. Although no one argues against this rule, surprisingly, it often is not followed. Many medical doctors seeing patients have many opportunities to find out about and treat sexual problems. But they often do not ask patients about this area of their health. Also, doctors often do not tell patients that a drug they prescribe might cause a sexual problem, or that a patient's medical illness may affect sexual function.

Protecting privacy. Therapists may find themselves in a "double bind" situation with patients, when no matter what the therapist does, he or she may feel helpless or guilty. The therapist has a duty to protect the patient's privacy and secrets. And patients and research subjects must be told that the information they give the therapist will be kept private. Most people will not talk to a therapist if they think the therapist is going to tell their secrets.

But therapists also have a duty to tell the truth to their patients. What happens if the therapist treats a couple, and the husband says he has had a homosexual affair? Should the therapist tell the wife? The therapist has an obligation to keep the man's secrets but also has an obligation to tell the truth to the wife.

Keeping a professional relationship. Because sex therapists are human, they may have trouble with their own sexual feelings during treatment. They may be attracted to their clients and have to struggle with these feelings. Most therapists follow the rule of ethics that advises all health professionals never to become

Studies indicate that 5 percent of psychiatrists have had sexual intercourse with patients. These rates are lower than those for general practitioners, surgeons, and gynecologists. In most cases, male physicians seduce women patients.

sexually involved with a client. If they do become involved, the relationship can be dangerous for both the therapist and client.

Special Issues in Therapy

Sexual contact between therapist and patient. Most therapists believe that sexual intimacy between therapist and patient should not be a part of sex therapy. They believe that it is bad for the professional relationship between therapist and client, and that it can cause great psychological harm to the patient.

Sexual surrogates. A sexual surrogate is a trained person who helps the patient with physical aspects of sexuality. The surrogate is not the therapist, but someone who works with the therapist.

Most sexual therapies involve exercises that the patient cannot do without a partner. Some therapists try to help people without partners by providing a sexual surrogate who engages in private sexual activity with the patient.

Critics of surrogacy. Some people criticize this, and say it is no different from prostitution. They also say the surrogates provide only physical training and do not deal with the emotional problems that often are part of poor sexual functioning.

People who use surrogates say that unlike prostitutes, the surrogates are screened, trained, and supervised. They also claim that people who do not have partners cannot get the treatment they need without surrogates, and that preventing them from getting this care is wrong.

In North America, this therapy is very controversial. In other countries, sexual surrogates are used more often. In Japan, people without partners can go to "sex school" and try different techniques with clothed models under a teacher's supervision.

Professionalization

So many people may need sex therapy that it is difficult to insist that only people with specialized training can be sex therapists. Masters and Johnson estimated that about half of all married couples will develop a sexual problem that could be helped by sex therapy.

Nonprofessional resources. There are others who help individuals with sexual problems. People may go to the hairdresser for a makeover, tell their problems to a bartender or friend, or visit a prostitute. Self-help books, articles in popular magazines, and "family life" classes in schools and churches are also available to help people. All of these nonprofessional resources will continue to be available. How should professional sex therapists work with people to make sure that they receive the right information and care? Should professionals help these other sources to provide accurate information, and get them to tell people when professional treatment is needed?

see also

Professional–Patient Issues: Professional–Patient Relationships

▲ A sexual surrogate in a session with a patient (1974). The therapy provided by a sexual surrogate has been very controversial in the United States.

Health-care team. Many health-care providers have become more involved in basic counseling and brief therapy for their patients. Also, treatment programs that involve many different health-care providers and one patient have become more common. Such programs may specialize in difficult or complicated cases. A man who has prostate surgery may have sexual difficulties after the surgery. His health-care team may involve a doctor, a nurse, and a social worker, and each of them may discuss sexual issues with the patient. All of these people must work together so that money is not wasted and the patient receives the right care.

Many health-insurance programs do not pay for sex therapy. This means that often only wealthy people receive sex therapy. It also makes it hard to tell if therapy is helpful or not, because in order to know this, therapists will have to treat many different classes of people, not just those with the money to pay.

Research

Many ethical questions come up when we consider research on sexual response, sexual problems, and sexual behavior. Most research projects have to be approved by reviewers who check to be sure that the research will not hurt the subjects. Research subjects should be protected from physical and psychological harm. They must be told about the risks and benefits of the research, so that they can choose for themselves whether or not they want to be involved. This choice should be their own, and they should not be forced or pressured to participate. Research subjects are also usually told that no one outside the project will know their identity.

In addition to these concerns, research should not affect the subject's emotional state, sexual adjustment, or relationships. A study may involve the showing of **erotic** photographs to the research subjects, but some subjects may not want to see them. Other studies may ask the subject to reveal very personal information. Another may involve observing or measuring sexual responses during sexual activity.

erotic tending to arouse sexual desire

Childhood sexuality. The use of children as research subjects is troublesome. Research on childhood sexuality could help people answer questions about the beginnings of sexual orientation, the formation of beliefs and attitudes about sex, and the development of unusual sexual desires. But children must be protected from sexual stimulation and emotional distress. Because of this, research about childhood sexuality usually involves adults remembering their childhood and telling researchers what they experienced. When children have been involved, research is limited to safe topics.

Research issues. It is difficult to decide what research is most important in a field that is so new. How should money and time be spent? Should researchers be involved in guiding public policy and health care? Almost no research has been done on the effect that various medical disorders and treatments have on women's sexuality. Should this research be done immediately? What about exploring the special needs of people with physical or emotional disabilities? When should we investigate **pedophilia**, incest, or violent sexual behavior? If we do not look at these areas, the social cost is great in terms of crime and human suffering.

Even apart from social issues, what fundamental questions about sexuality need to be answered? What causes a person to be heterosexual? Or homosexual? What are the connections between sex and eating, sex and pain, sex and danger, or sex and profound grief? What is the biochemistry of sexual attraction and arousal? And ultimately, how is sex related to love?

pedophilia a sexual perversion in which children are sexual objects

SEXUAL ETHICS

sexual orientation a person's understanding of the nature of his or her own sexual attractions, whether toward the opposite sex, the same sex, or both

artificial reproduction childbirth that is not caused by sexual intercourse between a man and a woman

The health of the human body is a matter of concern to the science of bioethics, and so the relationship between health and sex concerns bioethics, too. In a sense, bioethics includes sexual ethics—or at least, it includes some of its key questions: What does human sexuality mean? What are the causes of sexual attitudes, **orientations**, and activities? What are their effects?

Both bioethics and sexual ethics have questions, answers, and opinions about a great many matters related to sex: the physical acts themselves, the right of individuals to decide who and what they are, ideals of growth and fulfillment, and the importance of understanding that the meaning of sex and sexual activity—sexual rights and wrongs—are not really private matters but are what a society determines they are. Some issues that are particularly bioethical in nature are birth control, abortion, **artificial reproduction**, diseases related to sexual activity, unusual types of sexual attraction, male and female roles, sexual behavior of doctors and other medical personnel, sex research, and advice or medical help on sex-related matters. All of these issues concern bioethics because they involve questions of morality. When people speak to doctors and other health professionals on any of these matters, they are asking the health professionals to answer questions of sexual ethics.

Knowing the history of sexual ethics will help in understanding the ethical questions that have arisen about sex and people in the modern era. That is why the rest of this article will be about the many Western traditions of sexual ethics—traditions based in philosophy, science, medicine, and religion—and about the issues related to sex that many people living in the Western world find troubling.

Ancient Greece and Rome

The people of ancient Greece and Rome appear to have accepted sex as a natural part of life. Both societies permitted men to act

incest sexual relations between people who are closely related

bigamy the state of being married to two different people at the same time

adultery sexual relations that involve a married person and someone to whom he or she is not married

sexually in almost any way they wanted. Athens prohibited its male citizens only from engaging in **incest**, **bigamy**, and **adultery**. (Adultery was considered a crime against property, like theft: a Greek man's wife was his property.) But the attitudes of the two societies differed in important ways. The Greeks thought and wrote much about sexual relations between grown men and young teenage boys, while the Romans believed that the marriage of a man and a woman was the foundation of social life.

Marriage. For both Greeks and Romans, marriage was monogamous—that is, a man or woman could have only one spouse at a time. But in Greece no sexual ethic confined sex to marriage. Although male and female citizens were expected to marry, the basis of marriage was not a bond of love between husband and wife but the duty of citizens to preserve and maintain their society by producing children. The Greeks thought that men, by their nature, desired sexual relations with both women and other men and accepted these needs as natural; they also accepted the idea that men and women could live together without being married. Male and female prostitution and sexual relations with slaves were common as well.

In practice, much of this was also true in ancient Rome, even though ideals of marital faithfulness became much more important. Historians consider the development of marriage as a social institution a central achievement of Roman civilization. A major part of this development was a growing appreciation of the importance of the bond of love between husband and wife.

A young man plays the flute for a young male friend. Classical Greek culture valued friendship between men more than marriage, which was considered as a duty necessary to maintain property and nation.

Sexual morality. Greece and Rome were societies dominated by men, and the standards of sexual morality differed for the two sexes. Greek and Roman brides, but not bridegrooms, were expected to be virgins. In Greece, though wives were given the power and responsibility of managing the home, they were allowed no part in life outside the home. In Rome, even within the home, the husband's authority was final—he could rule with an entirely free hand. The ideal of male domination reaches its highest form here.

Both Greeks and Romans accepted **homosexuality**. The Greeks, especially, thought men were sexually attracted to all beautiful individuals, male or female. The idea of sexual desire interested them more than who the desired person was, or whether that person was a man or woman. Every Greek man was expected to marry in order to produce a child to inherit his property. But the Greeks thought that love and friendship between men—with or without sex—was a more precious thing than anything that marriage could provide, for men of all ages were thought of as equals.

The Platonic view. A group of Greek philosophers who lived in the sixth century B.C.E.—the Pythagoreans—believed in keeping the body pure in order to better develop the soul. Their thought influenced that of Socrates and Plato. Although Plato's writings do not express open disapproval of bodily pleasure, they do distinguish between higher and lower pleasures. In works called dialogues (because people are represented as discussing ideas with one another), such as the *Republic, Phaedo, Symposium,* and *Philebus,* Plato developed this opinion: Sexual pleasure is a lower pleasure and so ought to be controlled, since self-control is the goal for which all should aim. Plato thought that the great energy of physical love and desire should be directed at uniting the human spirit with the highest truth, goodness, and beauty. If bodily pleasure did not interfere with reaching this goal, he had no objection to it. But he thought that sexual intercourse did interfere with the human spirit's ability to focus on the greater good; he even thought that it reduced the level of tenderness and respect in individual relationships.

Roman philosophers of the late Empire. Many people believe that, during the last years of the Roman Empire, rules governing sexual conduct collapsed as part of a larger breakdown of morality in general. Modern scholars think, though, that those years witnessed a growing distrust of sexuality and a growing disapproval of various sexual practices and activities. Indeed, a distrust of sexual desire and a low opinion both of sexual pleasure and of everything having to do with the body (as opposed to the soul) are among the aspects of Roman thought that came to have the most influence on later Western ethical thinking. While sex was not considered evil, it was considered dangerous—and not only if there was too much of it. Some compared the orgasm to a seizure caused by the disease of epilepsy, while others thought a man's ejaculation of sperm would weaken him.

---

homosexuality sexual attraction or activity between persons of the same sex

see also

Sex and Gender: Homosexuality

The Stoics—a group of mainly Roman philosophers—probably had the greatest effect on later Western thought about sex. Some of them—Musonius Rufus, Epictetus, Seneca, Marcus Aurelius—believed that a person's willpower could control emotions and should do so for the sake of inner peace. They thought that sexual desire was much like fear and anger: something irrational that easily got out of hand. It should be kept under control or got rid of completely. It should be used only for a rational purpose, never for its own sake. **Procreation** was that purpose.

Overall, the later Greek and Roman views on sexual ethics—those, that is, that influenced Western thinking the most—are far more restrictive than those that were to be found in ancient Greece. Doubt about the value of sex and a belief that the sexual impulse needed to be controlled were the main themes of those later views.

procreation the production of offspring

The Jewish Tradition

The earliest Jewish moral codes were simple and not tied to beliefs about the nature of God or God's part in human affairs. Like other ancient codes from that part of the world (usually called the Near East), the Jewish codes included laws that regulated marriage and forbade rape, adultery, and prostitution. Unlike their neighbors, who held different religious views, the Jews believed in a God who is beyond sexuality but whose plan for creation makes marriage and human **fertility** holy and the subject of a religious duty. At the heart of the Jewish tradition of sexual morality is a religious command to marry, have children, and form a family that is headed by the husband. This structure may be called the patriarchal (fatherly) model for marriage and family. Many of the specific sexual rules stem from the duty to have children within this patriarchal structure.

fertility in humans, the condition or state of being able to have children

Religious Perspective: Judaism

Marriage. While having children was the principal reason for the command to marry, the Jews also believed that marriage helped the spouses become holy. They understood holiness as referring to the companionship and mutual aid that married people give each other; they also saw it as referring—secondarily—to the management of sexual desire. In fact, they saw monogamous lifelong marriage as the ideal setting for sexual activity, and in time the ideal became the custom.

Sexual morality. Because of the teaching in the book of Genesis (the first book of the Hebrew Bible), Judaism—that is, the religion of the Jewish people—has traditionally been concerned with what it regards as sinful or improper acts that involve male sexual ejaculation. These acts include **masturbation** and homosexuality. It considers homosexual acts unnatural, irresponsible (because they do not produce children), beneath human dignity, evil (because they represent uncontrolled sexual desire), and a threat to marriage and the patriarchal family.

masturbation sexual stimulation of a person's own genital organs

The Jewish tradition treats women's sexuality differently than men's. This came about in part because of women's subordinate role in the family and in society. Getting and keeping control over women's sexuality was necessary for a stable family, in the traditional Jewish view. Different rules applied to women and men about sex before and outside of marriage. There were even different rules about rape.

On the whole, the traditions of the various forms and branches of Judaism on sex and sexual activities have been positive, but not entirely so. The sexual instinct was considered a gift from God, but some teachers still called it an evil impulse. Different schools of Jewish thought had distinct opinions about the relationship between sacred and sexual matters.

Since the 1970s, many Jewish thinkers have expressed untraditional opinions about sex before marriage, birth control, abortion, the equality of men and women, and homosexuality. Others disagree about how to interpret traditional values in the modern world.

Early Christian Traditions

Like the teachings and traditions of other religious and cultural groups, Christian teachings about sex are complex and the product of many influences. They have also changed and developed over the course of many generations. The command to love God and neighbor—a command at the center of Christian moral life—comes directly from the teachings of Jesus and his followers as they are recorded in the New Testament. The New Testament also makes possible a sexual ethic that has three characteristics: It values not only marriage and procreation but also singleness and **celibacy**. It gives at least as much weight to internal matters (attitudes and thoughts) as to external ones (actions). And it gives sexual intercourse a sacred symbolic meaning, while at the same time treating it as a possible source of evil.

celibacy complete sexual abstinence; also, the state of not being married

▼ *The Marriage of the Virgin*, by Italian painter Domenico Ghirlandaio (1449–1494).

▲ The Holy Family was painted by Michelangelo in the early sixteenth century. It depicts the Trinity of Joseph, Mary, and the infant Christ in a joyous and carefree environment.

Historical background. Christianity came into the world at a time when Judaism was under the influence of various Greek and Roman philosophies and religious views. But unlike these sources of influence, Christianity's main concern was not self-control or the preservation of a society or an empire; and unlike the Judaism of that time, which was concerned with life in this world, Christianity was interested in the relation of life in this world to the life in a world to come. Both Christianity and Judaism tied their moral codes to their belief in God, and both viewed marriage and procreation as aspects of God's love and creative work. But like the Stoics, Christians distrusted the passions of the body. In the end, the Christian moral view taught that sex was good (because it was part of what God had created) but seriously flawed (because sexual passion is too strong to be controlled by reason). The Stoic view that linked sexual intercourse with the rational purpose of procreation made great sense to early Christian thinkers.

Christian sexual ethics adopted a course that did not change for centuries. The mainstream view was that sex was essentially good primarily because of procreation and despite some evil tendencies. This view was frequently challenged but always defended. The defenders of this view based their arguments in large part on the work of Saint Augustine (354–430 C E.), the great Christian religious thinker and writer, who took and developed ideas from both the Stoics and other Christian writers.

Saint Augustine. Against those who considered sex evil, Augustine argued in defense of marriage and procreation—even though he agreed with his opponents that sexual desire was an evil passion in itself. But he understood evil to be related to the absence of rational order. This disorder or inappropriateness was what made sex outside of marriage a destructive force, rather than a force that could contribute to love of God and neighbor. This positive reordering could be done only when sexual intercourse was confined within the marriage of a man and a woman, with the making of a child as its goal. Augustine believed that intercourse within marriage but without this purpose in mind was sinful, though not always seriously sinful. He taught that intercourse in marriage had a three-part purpose: to produce children, to promote faithfulness between husband and wife, and to keep the marriage bond firm and unbreakable.

Despite the fact that Augustine and most of those who followed him considered neither the body nor marriage evil, many in the Christian tradition for its first thousand years held a negative view of sex. This attitude was a product of the tone of many early Christian writers, who were, first and foremost, concerned about freeing themselves from desires that could not lead a person to God. In this struggle to free both the body and the spirit, even procreation did not seem very important. The ideal of celibacy took first place, ahead even of regulated sexual activity within the family.

Saint Thomas Aquinas. This thirteenth-century thinker was the next great figure in Christian sexual ethics, despite the fact that he wrote little about it. He taught that sexual desire was not evil in itself, since no naturally arising bodily or emotional inclination was evil in itself. An action became morally evil because of an evil moral choice. Because of people's tendency to do evil—a tendency that religious writers often call original sin—natural human inclinations tend to be disordered. Aquinas taught that though this disorder marked sexual passion, the passion itself was not evil. It became evil only when the disorder was the object of a moral choice.

 The Original Sin, a fresco at the cathedral of Gurk, Austria.

Aquinas explained in two ways why only sex linked with procreation was good. The first was to accept Augustine's view that sexual pleasure disrupts the working of the mind unless it is brought under rational control through a goal—procreation. The second was to argue that reason does not make up this goal: It discovers the goal by observing how the sexual organs of men and women are designed and function. Therefore, taking reason and rationality as guides to proper sexual conduct requires both the intention to procreate and the absence of any interference with the physical process of sexual intercourse (such as birth control).

Protestant Teachings on Sex

Questions about sexual behavior played an important role in the Protestant Reformation, a movement to reform Christianity that began in the early sixteenth century. For centuries, many priests had failed to observe the requirement that they remain celibate. Many Protestants thought that celibacy should not be a Christian

see also

Religious Perspectives: Protestantism

Martin Luther was so concerned about controlling sexual desire that he once said he preferred bigamy to adultery.

ideal. The reformers replaced celibacy with marriage and family as the center of Christian sexual ethics. Martin Luther and John Calvin were both deeply influenced by Augustine's teachings about original sin and its relation to human sexuality. Yet both developed a position on marriage that did not depend on procreation as an ethical guide. Like most of the Christian tradition, they taught that marriage and family were good, since they were part of the divine plan of creation. They thought, too, like Augustine, that original sin had caused humans to fall from grace, disordering the sex drive.

Lutheran sexual ethics. Luther was convinced that the cure for disordered sexual desire was marriage. He thought that sexual pleasure did not have to be explained or justified: the desire for sexual activity was simply a fact of life. Though there were elements in sex that drew a person away from the knowledge and worship of God, he argued that these sinful elements simply had to be forgiven, as did the sinful elements that appeared in all human activity.

Calvinism and sexuality. Calvin, too, saw marriage as a way to correct the disorder of sexual desires. He accepted the opinion that the married state was the best place to find and fulfill human happiness, but he went even further. He argued that what was best about marriage and sexuality was the intimacy formed between husband and wife. Calvin was more hopeful than Luther about the possibility of controlling sexual desire, though he, too, believed that any faults remaining in that desire were "covered over" by marriage and forgiven by God.

As part of their teaching on marriage, Luther and Calvin opposed sex outside of marriage, either before marriage or with a partner other than a person's own spouse. They also opposed homosexual relations. Both Luther and Calvin opposed divorce, but they agreed to allow it if a partner committed adultery or was **impotent**.

impotent physically unable to have sexual intercourse

Modern Roman Catholic Developments

During and after the Roman Catholic Counter-Reformation, from the late sixteenth century on, some Catholics offered new ideas and attitudes on sexual ethics, while others continued to argue for the ethic taught by Saint Augustine. The Council of Trent (1545–1563) was the first ecumenical (worldwide) council to consider the role played in marriage by love. It put love in second place to procreation and emphasized the superiority of celibacy to all other ways of life. In the nineteenth century, Catholic teaching on sexual ethics focused on "sins of impurity," a term that referred to the choice of any sexual pleasure unrelated to sexual intercourse within marriage and intended for procreation. In the twentieth century, Catholic thinking and teaching began to focus more on the person acting and less on the act itself. Protestant teaching of the same period considered birth control to be morally acceptable.

see also

Religious Perspectives: Roman Catholicism

encyclical a letter from the pope to all the bishops of the church or to those in one country

rhythm method a birth-control technique that involves refraining from sexual intercourse during the time period in which pregnancy is most likely to occur

Evolution of ethics. In 1930, after the Anglican Church in England gave its approval to birth-control devices, Pope Pius XI restated the traditional Roman Catholic sexual ethic of procreation in the **encyclical** letter *Casti connubii*. But he also gave approval to the use of the **rhythm method** for limiting procreation. Catholic moral and religious writers and thinkers began to think cautiously that sexual intercourse within marriage might be allowable solely to promote the marital union, even without the intention to procreate. The change in thinking among Roman Catholics from the 1950s to the 1970s was dramatic. The rhythm method had driven a wedge between procreation and sexual intercourse. This event, combined with new ideas about the wholeness of the human person, resulted in books and articles that spoke of sex as an expression and cause of married love.

Contemporary issues. Pope Paul VI, in his 1968 encyclical letter *Humanae vitae*, insisted that contraception was immoral. But instead of settling the issue, the letter started a conflict among Roman Catholic thinkers and writers. Most disagreed with the pope's teaching, though sometimes for different reasons. Catholics continue to ask serious questions about specific moral rules on sexuality, and experts continue to debate sex outside marriage, homosexual acts, remarriage after divorce, treatments for infertility, the nature of men's and women's roles, and the requirement that priests remain celibate.

Post-Reformation Protestantism

After the Reformation, Protestant religious thinkers and church leaders continued to argue that heterosexual marriage was the only acceptable place for sexual activity. Except for somewhat different views on celibacy and divorce, Protestant sexual ethics looked much like those in the Catholic tradition. But twentieth-century Protestant ideas about sexual ethics changed even more dramatically than did those of Roman Catholics.

Contraception. The fact that Protestantism in general had always depended less on procreation as the basis of its sexual ethics made it possible for almost all Protestants to think of contraception as morally acceptable. By the 1990s, mainstream Protestant sexual ethics no longer agreed with the idea that sexual desire was immoral, self-centered, or dangerous.

Alternative lifestyles. But disagreements remained about authority within the family and about so-called alternative lifestyles: unmarried heterosexuals living together and sexual partnerships and activity between homosexuals, for instance. Though most Protestant thinkers agree that marriage is still the best context, or place, for sexual intercourse, most also are willing to accept the idea of sex outside of marriage for both heterosexuals and homosexuals, so long as the persons involved are acting of their own free will.

In 1780, the moral philosopher Immanuel Kant wrote that through sexual appetite one human being will often plunge another into "the depths of misery," casting him or her aside "as one casts away a lemon which has been sucked dry." Kant valued sexual intimacy, but only within the context of lasting commitment, mutual respect, and love.

psychoanalysis a method of treating emotional and mental disorders that involves exploration of a person's memories and dreams, especially those from early childhood

In his book *Rediscovering Love* (1986), psychiatrist Willard Gaylin summed up the results of the sexual revolution of the 1960s and 1970s: "the spread of two sexually transmitted diseases, genital herpes and AIDS; an extraordinary rise in the incidence of cancer of the cervix; and a disastrous epidemic of teenage pregnancies."

In her book *The Divorce Culture* (1998), Barbara Defoe Whitehead describes the change from an American culture in which marriage was expected to last to one in which it is not. She hopes that our cultural institutions can bring about a new culture in which marriage commitments are taken more seriously for the benefit of both spouses and children.

Philosophical Developments

Philosophy and philosophers have paid little attention to sex throughout history. Though they have written much about love, they have left sex largely to religion, poetry, medicine, or the law.

Freud and Psychoanalysis

When psychoanalytic theory began to appear in the early twentieth century, it offered new ideas about the meaning and role of sexuality in people's lives. Whether the views of Sigmund Freud turn out to be valid or not, they have had a tremendous effect on traditional sexual morality. Freud agreed with Augustine's and Luther's ideas about the nearly uncontrollable nature of sexual desire, but he saw sexual need as a natural drive, not the result of sin, and, more important, as the center of the human personality. Freud described earlier efforts to guide or control sexuality by means of reason and rational goals as repression. After Freud, many people came to regard problems involving sex as problems of mental, or psychological, illness, not as moral problems. Medical treatment, not forgiveness, became the path to freedom.

Psychoanalytic theory has raised as many questions as it has answered. Though Freud wanted people to be free of the idea that anything concerning sex should be forbidden, or taboo, he also believed in sexual restraint. A concern for ethical guidelines thus remained, and Freud's own recommendations were in many ways quite traditional.

Other Influences

In the social and scientific thought of the nineteenth and twentieth centuries, Freud's opinions were not the only ones that reshaped Western sexual standards. Biologists who studied human sexual reproduction offered new ideas on male and female roles in sex and procreation. Anthropologists discovered that people who came from different cultural groups had different sexual ethics, making the idea of a single and unchanging human nature a very doubtful one. Surveys were taken about the kinds of sexual behavior in which people took part. Many of the surveys showed that there was a great gap between accepted standards of sexual conduct and what people actually did.

Rebuilding Sexual Ethics in the Twenty-first Century

Developments in the ways teachers and scholars think about historical matters and about the study of history itself have clearly weakened traditional sexual standards. Writers and thinkers on sexual ethics are as much at sea as anyone else. These writers and thinkers find that science and medicine sometimes help in the search for ethical standards, but sometimes do more harm than good. The traditions that have their roots in philosophy and religion are also searching for answers that people can use, as well as for clear, reasonable, and inspired guidance.

Those who concern themselves with sexual ethics in the modern era see many meanings in human sexuality—pleasure, reproduction, love, communication, conflict, and stability among them. They wish to guide sexual behavior in ways that will increase its good possibilities and restrict its evil ones. They see certain parts of sexual relationships—safety, nonviolence, equality, independence, truthfulness, and willingness on all sides to be involved—as essential. Many think that care, responsibility, commitment, love, and fidelity are required, too, or should at least be goals.

SEXUAL IDENTITY

The term "sexual identity" is controversial and difficult to define The traditional definition is that it refers to whether a person is biologically male or biologically female. Sometimes people who take this traditional position also believe some other things about sexual identity. They may believe that males are naturally "masculine" and that females are naturally "feminine." They may believe that males and females are naturally sexually attracted to each other and that this attraction is most naturally expressed through sexual intercourse. They may also believe that although people want to have intercourse because they are attracted to each other, they should also have other reasons, the most important being love and the desire to be married and have children.

In the last half of the twentieth century, various groups began to question these ideas about sexual identity, male and female roles, and the place of sex in human life. These groups included **feminists**, people who believed in sexual freedom, gays and lesbians, and scholars. All of these groups raised questions about what sexual identity is. Some of their questions include:

feminist someone who supports the political, economic, and social equality of the sexes

1. The sex question: Are there really two distinct biological types, "male" and "female"? Is there always a clear difference between them?

2. The gender question: How should we think about the relationship between biology and psychology? If someone is male, does that mean he will always have particular "masculine" traits and behaviors?

3. The sexuality question: What is sexual desire? How can it be described?

4. The sexual ethics question: How should we think about sexual acts? Which acts should be considered right and which considered wrong? What makes a sexual act right or wrong?

Sex

In the late twentieth century, many people began to question the belief that some people are all male, some are all female, and no one is in between. In this view, each sex is associated with certain physical characteristics. Most people believe that men are bigger and stronger than women, but feminists claim that this view is limiting and untrue. They say that this belief is shaped by society, which historically has emphasized the differences between men and women. Instead of looking for these differences, feminists say, we should assume that any physical characteristic could show up in either a man or a woman. Although generally men are "bigger" than women, it is easy to look around and see individual women who are taller, heavier, have longer legs and arms, or are stronger than many men. Instead of a strict division between male and female physical traits, there is a lot of overlap.

In the same way, most people think that women and men have very different **hormones**. For example, estrogen is considered the "female" hormone and androgen is considered the "male" hormone. This suggests that women have one and men have the other. But in reality, both hormones are found in both women and men. After **menopause**, women often have a higher ratio of androgen to estrogen than men who are the same age.

People who study the history of human culture claim that the view of men and women as having very different body types is relatively new, and has developed only within the last few centuries. Before the eighteenth century, people thought that women's bodies were less developed versions of men's bodies.

During the eighteenth century, people began to believe that female and male bodies were basically different. They began to give different names to sexual organs that had been given the same name before. Even parts of the body that had nothing to do with reproduction, like the skeleton and the nervous system, began to be thought of as very different in women and in men.

Biological research in the late twentieth century has suggested that it is not always easy to tell a "male" from a "female." There are many traits that can show whether someone is male or female, but they are not always clear-cut. Even when they are, they sometimes do not definitely determine whether someone is male or female.

Most scientists have used **chromosomes** as a marker to determine whether someone is male or female. Chromosomes transmit genetic information from the parent to the child. Most people have two sets of chromosomes, one from each parent. Generally, females

hormone a substance secreted by a cell, tissue, or organ, which circulates in the body and stimulates functions in certain other cells, tissues, or organs

menopause the natural termination of menstruation, occurring usually between the ages of forty-five and fifty-five

chromosome a rod-shaped body that contains genes of an organism; different organisms have differing numbers of chromosomes to hold all the genes

have two X chromosomes, and males have one X and one Y chromosome. This seems like a clear distinction, but it is not always so. Some people inherit only one X chromosome, and no Y chromosome. Or a piece of a Y chromosome may become attached to an X chromosome, producing a person with an XXY pattern.

Even individuals who have a standard XX ("female") or XY ("male") pattern may have characteristics that would lead people not to see them as female or male. An XY person may not be able to produce the "male" hormone testosterone, or may have cells that are not sensitive to testosterone. That person will end up looking more like a female than a male.

Sexuality

At least since the 1890s, in industrialized Western countries, people have believed that one kind of sexuality was "normal." This involves intercourse between one male and one female. All other sexual practices have been seen as "abnormal" or "perverse."

Homosexuality has been the subject of a great deal of debate and controversy. Since the late nineteenth century "homosexual" has been used to label some people in a negative way. Homosexuality has been against the law in many places, and people have been put in prison for homosexual acts.

It is not clear whether people are homosexual because they are born that way or whether they become homosexual as they grow up. During the 1960s and 1970s, homosexual men, who began calling themselves "gay," and homosexual women, who preferred to be called "lesbian," began to form political alliances. These groups worked to resist the laws, practices, and beliefs against them. They stated that homosexuality is not a perversion or a sickness, and that it should not be outlawed and did not need to be "cured." They said it was simply a difference in sexual orientation that should be tolerated in a free and open society.

Other questions have been added to the debate about homosexuality. Does "homosexual" describe a particular type of person, or should the word be used simply for a specific type of activity that anyone might engage in? People did not always have the concept that "the homosexual" was a particular type of person. This concept is relatively new and began to be used in the late nineteenth century. At the same time, people began to use the word "heterosexual" for someone who is sexually involved with the opposite sex.

People did not always believe that individuals should be defined by the sex of their partners. In many Native American societies, certain men, called berdaches, took on many of the tasks and roles that women usually did. These men would have sex with other men. However, they were considered different not because they had

Oscar Wilde (1854–1900), Irish poet and dramatist, was tried and convicted for being a homosexual in Victorian England.

sex with other men but because they took the passive role. Their partners were not considered any different from men who had sex with women.

The way we see sexual practices defines much of how we think about sexuality. The belief that "normal" sex is always heterosexual makes it seem as though sex between people of the same sex is abnormal. These categories of normal and abnormal also generally have strongly moral overtones. For this reason, people who believe that homosexuality is abnormal also believe that it is morally wrong. The idea that homosexuality is normal sex, that homosexuals are not sick or evil, is widely accepted in many cultures.

Sex and Gender: Homosexuality

Sexual Models

Although people have often seen sexual identity as a simple matter—one is either male or female—it is not always that clear. In the same way, people have often tried to simplify the many varieties of sexual behavior—either a behavior is "normal" or "abnormal."

Ancient Greek thinkers, such as Plato and Aristophanes, believed matters were not quite so simple. One famous Greek tale is that each human being was once part of a divine whole, which the gods split in half. Ever since, each person has looked for his or her other half, and sexual desire is simply this deep longing to join with the partner and become whole again. The philosopher Socrates had a somewhat more superficial view that sexual desire was the love of beauty.

The reproductive model. Biologically, sexuality can be specifically defined as part of human reproduction. This biological, reproductive concept is very important to our thinking about sexuality. No matter what variations in sexual activity humans experience, the fact remains that heterosexual intercourse is deeply connected with our survival as a species. Because of this, the idea of heterosexual intercourse can be rejected or argued with, but it cannot be ignored.

The view that sexuality and sexual desire are aimed at reproduction, even if the people involved are not planning to reproduce, is a more "murky" view of sexuality. Those who hold this view believe that even if people involved in sexual activity are just looking for pleasure, they are drawn together for deeper reasons, one of which is the drive to reproduce the human species. If people are religious, they may believe that God has made people with this drive so that, as the Bible says, they can "be fruitful and multiply."

All "murky" views of sexuality believe that sexuality has a purpose, or more than one purpose, and that we may not even know what these purposes are. Minimalists believe sex is a simple physical act, the desire for contact with another person's body. Murky views insist that sex means much more than this.

Pamphlets promoting sexual abstinence.

The pleasure model. Most minimalists are bothered by the restrictions on sex that are a part of Western Christian culture. For two thousand years, Christian theological tradition and some parts of the Bible have insisted that the main, or only, purpose of sex is to produce children. This tradition does not value the pleasures and desires associated with sexual activity. In this tradition, emphasizing pleasure instead of reproduction is considered wrong. For this reason, it has often been considered wrong to use birth control or to engage in sexual activity that cannot lead to reproduction.

The pleasure model opposes the reproductive model, with its restrictions on sexual activity and its belief in a deep human drive to reproduce. Since the 1960s, many birth control methods have been available. The resulting ability to have intercourse without having to reproduce has made the view of "sex for pleasure" more attractive to many people. The idea that sex is mainly for pleasure now seems obvious.

People who believe sex is for pleasure place fewer restrictions on it than people who follow the reproductive model. Although many people prefer heterosexual intercourse, it is not considered more normal than homosexual intercourse or other acts. In this model, sex is "good" if it provides pleasure to those engaging in it and "bad" or mediocre if it is not satisfying.

Although there are fewer restrictions on sexual activity in this model, it does not mean that people can do whatever they want. Rape is still wrong. And while almost any activity that adults agree to is acceptable, forcing someone to be sexual when they do not want to be is not acceptable. It is considered normal for people to use birth control, and sex for reproduction becomes less common. The appeal of this model and most contemporary sexual ethics is that sex should be pleasurable and, except for a few moral limits, should not be restricted.

Some people might call the pleasure model the "Freudian" model of sexuality, after Sigmund Freud, on whose work it is based. In his book *Three Contributions to the Theory of Sex* (1962), Freud said that sex should be seen as enjoyable in itself, and should not be judged by whether or not it leads to reproduction or other ends.

Even though the pleasure model seems simple, Freud is not simple. He was one of the main thinkers who showed how deep and complicated the human mind is and how complex is human sexuality. Sexuality and pleasure are not simple or obvious. Over two thousand years ago, the Greek philosopher Aristotle noted that pleasure is not just a sensation. Pleasure comes from successful activity. It is not the same thing as satisfaction, but it comes with satisfaction.

The pleasure model invites different understandings of sexuality and satisfaction. What is it that is enjoyed? What is satisfaction?

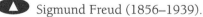 Sigmund Freud (1856–1939).

Why is something enjoyed? The same sensation of being touched can feel very different if the person doing the touching is a lover or a stranger. This makes it clear that pleasure is not simple but is connected to motives, thoughts, and feelings that are often hidden and not well understood.

The metaphysical model. Some of these motives are so deep that they could be called spiritual or philosophical, because they go beyond easily understood physical attraction and emotional desires. Sigmund Freud and another psychologist, Carl Jung, often wrote about these "metaphysical" aspects of sexuality.

One of the most basic metaphysical models of sexuality goes back to the fable in the Greek philosopher Plato's *Symposium*. This fable tells the story of how each human being once existed as one half of a unified whole, which the gods split and made into two people. In the fable, sexual desire is the longing of the two halves to be one again. This story obviously is not literally true but shows a deep insight into the feeling of "merged selves" or oneness that people may experience with their sexual partner.

In the metaphysical model, people have a deep desire to be together, and sexual activity is an expression of that desire. This concept is related to the idea of romantic love and the idea that two people were made for each other.

Even though the pleasure model of sexuality is common in modern culture, most people want more than just pleasure from sexual activity. They want deep or meaningful relationships. People who follow the metaphysical model reject sex without love, even if it is pleasurable. In this model, sex has to be part of a loving relationship.

The communication model. What about sexual activity that is not aimed at reproduction, pleasure, or romantic love and togetherness? What about sexual activity that seems based on one person inflicting pain on the other? What about sexual activity that may be tender, but is not a part of a loving relationship?

Sex can be seen as a kind of communication. It can be a physical form of expression of a person's emotions and attitudes toward other people. It is a language, but it is very different from spoken or written language. Instead of words, there are touches, gestures, and physical positions. It may be an expression of domination or of submission. It may express respect, fear, tenderness, anger, admiration, worship, or concern, as well as love.

The communication model shifts the emphasis in sexuality away from the physical and sensory aspects of reproduction and pleasure to the interpersonal connection between people. Instead of expressions of love, it emphasizes expression of all emotions and attitudes. All of the models recognize that sexual partners share their emotions, but the communication model believes that this sharing is the most important aspect of sexual activity.

The Hindu *Manual of Love (Kama Sutra)* was written in the fifth century C.E. It is a categorical organization of every aspect of love, courtship, and marriage. The *Kama Sutra* enjoins males to complete their education and then to marry and set up a home. Both men and women are encouraged to study and to perfect the methods of love to give their lives a greater esthetic quality. This text supports the idea that the experience of physical love between two people of either the same or opposite sex is to be engaged in and enjoyed for its own sake.

▲ Jean-Paul Sartre (1905–1980).

Existentialism. In the 1940s, the French philosopher Jean-Paul Sartre proposed a form of this model in his work *Being and Nothingness.* He understood all sexuality to be an expression of conflict, a war for domination and freedom. But sex does not involve only these feelings. Yet even though Sartre's position is extreme, he forces us to see something that is left out in the pleasure and the metaphysical models. Sexual relationships are not always innocent and loving, even when they are normal, willing relationships between people. Sex can be used to express all kinds of emotions, and some of them may not always be positive.

The models compared. It is clear that answering questions like "What is normal sex?" and "What is perverse?" is very complicated. In a strict reproduction model of sexuality, normal sex is only what leads to the conception of a child. Everything else either does not matter or is wrong. In fact, people who follow the reproduction model usually say that it is an important part of marriage and that love and pleasure are also important parts of sex.

In the pleasure model, by contrast, whatever gives pleasure to two adults who agree to do it, is normal and acceptable. In this model, a perversion is anything that provides pain instead of pleasure, that ignores the pleasure of the other person, or that produces pleasure in a way that is harmful in the long run.

In the metaphysical model, normal sex is an expression of shared love. In the communication model, it is complicated to decide what is normal, because one has to look at all the emotions being expressed and at both people involved to make a judgment.

Conclusion: The Problem of Normality

If we use biological traits to define sexuality, "normal" sexuality seems to be easy to define. Males and females have a different anatomy, and "normal" sex is intercourse between male and female. But as we learn more, we find that biology is actually very complex. In deciding what is normal based on chromosomes, the distinction between male and female becomes more and more difficult. And if we consider the fact that different cultures have very different ideas about what "normal" sexuality is, we can only get more confused.

How does one define what is "normal" and what is not? Some people would say that "normal" is what most people do, and many would insist that this is a good definition. But if we are talking about ethics, saying that something is "normal" also means that it is morally right.

When we look at sexuality, it is hard to decide what is "normal" by looking at what most people do, because sexuality is very private and involves many differences between people. No one really knows what most people do.

Of the various case models presented here, not a single one would be accepted as "normal" in every society and by everyone. Moreover, a pure example of an ideal type or model is probably nowhere to be found.

Whose standards do we use? Traditional religious standards? The more modern attitude that "anything is all right if it's between adults who agree to do it"? The current legal standard in the courts, which is that "local standards" should be used?

These complex questions should not lead us to believe that there cannot be any ethical norms in sexuality, or that there can be no such thing as normality or perversion. Instead, it is clear that these issues are very complicated, and it is best to tolerate the lives and decisions of others. Medicine, biology, social psychology, and anthropology are not the only fields that will help us understand sexual identity. We must welcome and accept different people and different views, and try to understand and appreciate them.

Related Literature

Sherwood Anderson's short story "I Want to Know Why" (1921) is narrated by a fifteen-year-old who is looking back on an experience he had the previous summer and is trying to figure it out. He was at a race track, very much enthralled with a particular horse, Sunstreak, and his trainer, Jerry Tillford. The teenager's emotions get all mixed up, feeling a kind of ideal love for the animal and for the trainer. But then he follows Jerry Tillford to a house of prostitution, and sees his hero looking at an ugly, mean-looking woman the same way he had looked at his horse. The narrator feels betrayed and very angry.

WOMEN

This entry consists of three articles explaining various aspects of this topic:

Historical and Cross-Cultural Perspectives
Health-Care Issues
Research Issues

HISTORICAL AND CROSS-CULTURAL PERSPECTIVES

A central problem of women's history is that women have been defined by men using concepts and terms that are based on men's experiences. "Women's studies" (a specialty only since the late 1960s) has made us aware that attitudes, customs, laws, and institutions affecting women have traditionally been based on ideas that include the following:

1. "Woman" was created from and after man.

2. Woman has been defined by her sexuality and her sexual function.

3. Woman has been confined to roles that are extensions of her ability to bear children.

For two centuries, **feminists** have criticized religion, philosophy, and science, as well as the family and political institutions, for assumptions that demean and harm the world's women.

feminist someone who supports the political, economic, and social equality of the sexes

Women Defined

From ancient times it has been customary to define "woman," in relation to man, as a limited and dependent part of a two-sex species. With little political power, women have been subject to countless laws and customs that both created and enforced their status as secondary beings. Men have defined women mainly in terms of their sexuality. At the same time, as masters of family and political power, men framed and staffed the institutions that control female sexuality. In the early fifteenth century, the French author Christine de Pizan (1364–c. 1430) challenged the then popular idea that women were evil by nature. She declared that this evil existed only in men's minds and that women would become as virtuous and capable as men if they were permitted education.

The rights of women. There is a long history of resistance and rebellion by individual women. Organized protest by women, termed "feminism" only since the 1890s, dates from the late eighteenth century. But treatment of women was only occasionally viewed by most people as a general problem of social justice. The "woman question," as it was called in the nineteenth century, was debated as a political, social, and economic issue, but rarely as a moral issue.

In the great democratic revolutions of the eighteenth century, the "inalienable rights of man" were not extended to women. Men, as heads of traditional families, continued to speak for their dependents, women as well as children. Ultimately, most improvements in the condition and status of women came about because they would prove beneficial to men. In 1915, feminists argued that if women voted, war would be less likely. In 1985, women argued that if mothers earned fathers' wages, fewer children would live in poverty.

Identity issues. Most matters relating to women, then, have been decided based on what would be most useful to society as a whole, and to men in particular. Woman as an individual person was almost never the subject of thought or decision. Woman was considered as a wife and mother or potential mother. Even the feminist leaders of the nineteenth and twentieth centuries resorted to the argument that giving women rights was the easiest or most useful

In the United States, women could not vote until 1920.

thing to do, rather than the most ethical or just. By the 1990s, however, feminist writers began to question such arguments and to demand a voice in establishing fields of knowledge and social institutions.

Women have been, and remain, deeply divided over their own definition of self. Are they individuals who are equal to and should be treated in the exact same manner as men? Or are they persons with gender-specific differences and resulting relationships and responsibilities who should be treated on their own terms? Awareness of the "equality versus difference" debate is critical to understanding a wide range of historical and modern-day women's issues.

If there is a failure to emphasize women's differences from men, women may continue to be viewed through a single, male-constructed lens that ignores or puts down female-specific experiences. "Woman" should not be defined only in relation to "man," just as all humans should not be defined by the experiences of white Anglo-Saxon men. Modern-day scholars recognize that neither "man" nor "woman" has a single, fixed meaning across all cultures and nations.

▲ A female construction worker.

Women in Traditional Western Societies

Studies suggest that society may have considered women more equal to men in prehistoric times. Farming economies that had few individualized tasks allowed for more equal relationships within families. Families themselves made up societies, and participation was not decided by gender or sex roles. As economies advanced and obtained greater wealth and military power, property rights and citizenship were given only to heads of households. Society became a male world, and women were confined to the household and considered men's property. A double standard of sexuality ensured that women were subjected to male interests. A woman's honor, and her family's, depended on her chastity. The Greek philosopher Aristotle said that the virtue of a woman was to obey. Women's roles in some part depended on their class, but no women could claim treatment equal to men with regard to property, citizenship, marriage, criminal law, or access to social institutions.

Influence of Christianity. The spread of Christianity initially brought new possibilities for women: for some, a role in spreading the new religion; for all, a promise of spiritual equality. Christianity created new opportunities for women's voices to be heard, especially by instituting marriage laws requiring consent and establishing, in some instances, inheritance and property rights for women.

The Middle Ages saw the founding of the first universities in the Western world, beginning in 1088 with the University of Bologna. Bologna's famous twelfth-century legal scholar Gratian developed a view (based on Aristotle's) that men were active and women were

The *Malleus Maleficarum* ("Hammer Against Witches"), a handbook of late medieval inquisitors, states that only women can be witches because their irrational and unruly passions lead them to physical intimacies with the devil.

midwife a woman trained to help other women in childbirth

Arthur Miller's play *The Crucible* deals with the witch trials in Salem, Massachusetts, in 1692.

misogynistic hateful toward women

passive—in law as well as in reproduction. In the thirteenth century, Thomas Aquinas also advanced this theory. Aquinas combined his reading of Aristotle with the Christian view of creation to assert that woman was a "defective and misbegotten" man, assigned by nature to bear children. Despite the rebirth of learning, women continued to be defined as "not men" (for the qualities they lacked) and "for men" (for the services they provided).

Neither the Renaissance nor the Reformation, both landmarks in European history, led to improved views toward women. The Protestant movement resulted in the closing of convents that had allowed some women an alternative to marriage. It also deprived women of the comfort of the Virgin Mary and female saints.

Urban versus rural experience. Controversial debate over the effects of the Renaissance and Reformation on women continues among historians. Some women prospered in these periods, enjoying education by leading humanist scholars and wielding power on behalf of dynasties. Urban craftsmen's wives shared in the production and marketing of goods, and helped to manage workshops. City women developed professions of their own, largely as healers, **midwives**, and store clerks. But most wage-earning women worked as household servants, often until marriage and sometimes for their entire lives. The term "maid" came to mean the same thing as "female servant."

Most women, like most men, lived in rural settings. There, all members of the household pooled their labor in order to produce the goods and services necessary to the economic survival of their families. Labor needs of the family determined the status, residence, and welfare of most people. Only after the centuries-long changes in agriculture and industry, along with reduced birth and death rates, did the employment of women generate debate over the "woman question." Ultimately, it was a shift in work location from the household to the factory and the subsequent change in women's economic contribution to the family that laid the foundation for feminist debate.

Effects of political and scientific developments. As political structures reorganized and science advanced, women's lives changed. According to one interpretation, the terrifying witchcraft persecutions of the sixteenth and seventeenth centuries reflected both religious and gender conflict as well as efforts by political institutions to control individual behavior. Because women were thought to be evil and open to the devil's temptation, and because it was believed that they would resort to magic to make up for their lack of physical and economic power, most witch-hunt victims were women. Often, these victims were older, single, eccentric women who lacked male protection.

Ultimately, science disproved many **misogynistic** ideas about the female body. Despite studies that challenged Aristotle's view that women were passive in reproduction, only in the late nineteenth

▶ Three witches burned at the stake by male executioners in a 1555 German broadside. The devil, depicted as a flying demon, watches the execution.

and early twentieth centuries were such false "classical" assumptions finally replaced by scientific knowledge.

Although by the eighteenth century the economic, political, and intellectual structures that maintained traditional attitudes and practices toward women were challenged, ancient patterns persisted. The Napoleonic Code of 1804, and similar codes of law passed across Europe, required married women to obey their husbands. Men alone wrote and signed the new "social contract"; as "natural" dependents, women could not try to obtain citizenship.

Women in Changing Societies

Inspired by the French Revolution, women in the nineteenth century formed groups that advocated improved treatment of their sex. By the mid-nineteenth century, organized "feminist" groups in France, England, the United States, Prussia, and even Russia challenged women's inferior status. Radical changes in agriculture and transportation methods, the rise of the market economy, the increase in industry, and the growth of cities brought important changes to family structures and relationships.

Susan B. Anthony

Susan B. Anthony (1820–1906) was a fighter for abolition of slavery, equal wages for equal work, and the of right of women to vote. She also organized The Daughters of Temperance, a group that fought alcoholism, and tried to change laws that made it impossible for women to own property.

Susan B. Anthony traveled throughout the United States and Europe to advance the cause of women, and by the end of her life she saw the feminist movement finally gain respectability.

Equal Rights Amendment (ERA)

An amendment to the U.S. Constitution for the recognition and enactment of equal rights for woman was introduced in the House in 1923 by Daniel R. Anthony, the nephew of suffragist Susan B. Anthony. Simultaneously, a parallel bill was introduced in the Senate. Drafted by Alice Paul of the National Woman's Party, the amendment read: "Men and women shall have equal rights throughout the United States and every place subject to its jurisdiction."

The first ERA instantly became a point of disagreement among women activists over the issue of special and protective labor laws for women. All attempts at compromise failed, and no action was taken until World War II brought back the issue of women's work in light of their contribution in the war effort. In May 1943, the Senate Judiciary Committee approved an ERA that stated: "Equality of rights under the law shall not be denied or abridged by the United States, or any State, on account of sex."

The Senate, voting first, passed the amendment by a narrow majority but failed to produce the two-thirds majority needed to send the amendment to the states for ratification.

Opponents to the ERA introduced a rider to the amendment stipulating that its first provision (equality) "shall not be construed to impair any rights, benefits, or exemptions. . . conferred by law upon members of the female sex." With this rider, which the Senate approved, the legislators voted for both equality for women and special treatment. The Senate sponsors withdrew their resolution, while in the House, arguments within the Judiciary Committee kept the amendment from being brought to general discussion.

By the 1970s, the labor laws that existed for women only were eliminated as a result of the ban against discrimination passed in 1964, and the women's movement was quick to unite around the need to achieve equality for women through a constitutional amendment.

After a two-year campaign during which women's organizations flooded Congress members with mail and telephone calls, both chambers of Congress passed the original ERA, sending it to the states with a seven-year deadline for ratification.

In spite of promising early results for its nationwide ratification, ERA ran into powerful opposition. The Supreme Court's decision in *Roe* v. *Wade* (1973) legalizing abortion provided conservatives with the arguments that abortion and ERA were linked, that the amendment would legalize homosexual marriage, eliminate support laws, and require the military to use women in combat. Supporters of ERA, still seeking three state approvals to claim victory, won an extension of the ratification deadline.

On June 30, 1982, the deadline arrived, but ERA had failed to garner the approval of the three-fourths majority of the states, as required by the Constitution.

For women especially, escape from male-dominated families led to new vulnerabilities. With women's wages far below necessary levels, an unwed single woman needed assistance, and might trade sex for survival, risking being fired for "loose morals" or being deserted by her male partner.

Social reformers responded to this situation, supposedly in women's defense. Not all protesters and reformers called for "equality" for women. Few, if any, considered identical rights and responsibilities for both sexes. Debates over the status of women most often focused on ways to "protect" them: to shelter traditional women's work from the intrusion of men; to keep women and children from unsafe work conditions or long hours; and to give women the rights to inherit property and keep their own earnings, to make decisions concerning themselves, and to share in legal authority over children in cases of divorce.

Defining feminism. Historians' focus on the women's **suffrage** movement (which began in the mid-nineteenth century and peaked at the beginning of the twentieth century) has obscured both the larger concerns of women activists and deep differences within feminist movements.

In earlier eras, as in modern-day developing countries, feminism was more closely associated with issues of class and nation and with family relationships and community ties. This makes up a "relational" form of feminism. Socialist feminists, while mindful of the need for women to receive education and encouragement in order to participate in class struggles, refused to believe that "equality" for women meant equal access to the conditions of working-class men, which they viewed as exploitative. Others, including many Catholic feminists, insisted on improvement of women's status in order to enhance their traditional roles and relationships. In some countries, notably the United States, a "century of struggle" for women's rights grew out of religious sentiment and the recognition that no person under another's power, whether woman or slave, could be fully responsible to God as a moral being.

Equal but different. Must arguments supporting a political movement on behalf of women—the various forms of feminism—be based on the assumption that human beings are identical? If so, equal-rights laws can be used to deny pregnant women special insurance and employment benefits. "Equality" between men and women may demand identical treatment.

Alternatively, to emphasize women's specialness, to focus on sexual differences, may lead to new laws that restrict women's options while pretending to acknowledge their special needs. Such an approach was used to deny many excellent employment opportunities to women because they required occasional evening work or involved physically demanding tasks. This "difference versus equality" debate has led to extended conflict over definitions of feminism and feminist demands.

suffrage the right to vote

Important Feminist Philosophers

Mary Wollstonecraft (1759–1797) was far ahead of her times when she wrote the *Vindication of the Rights of Woman* in 1792. The book has had a tremendous influence, even though scholars at the time did not take it seriously.

Simone de Beauvoir (1908–1986) followed Wollstonecraft in arguing that gender inequality is a matter of upbringing. Girls are taught that they are less intelligent and less talented than boys and are best suited for motherhood. In her book, *The Second Sex* (1949–1950), de Beauvoir urges equality, and in her own life she decided not to have children in order to pursue this equality.

Other important twentieth–century feminist thinkers include Germaine Greer and Betty Friedan.

The history of women in the twentieth century shows the importance of the "woman question" to the social, economic, and political concerns of many nations. During wars and revolutions, traditional ideas of "women's place" and struggles over women's suffrage have been overshadowed by the need for female labor and patriotic support. Apparent feminist advances during these times, however, have

▶ Women at parts factory during World War II.

The American feminist Betty Friedan (1921–) published *The Feminine Mystique* in 1963. In this book, she attacked the idea that women could find fulfillment only through childbirth. In 1966, she founded the National Organization for Women (NOW).

frequently disappeared after the need for women was over. Following both world wars, women were dismissed from well-paying jobs or offered less skilled and less rewarding employment.

Yet structural changes in business and industry have created greater demand for women workers, especially in clerical and teaching jobs. Meanwhile, expansion of educational opportunities has increased female literacy and professional expertise. Advances in public health, nutrition, and medicine have continued to increase women's life expectancy and decrease infant mortality. New technologies have reduced the need for time-consuming household chores. All of these changes free many women for long periods of productive activity outside the family.

The rebirth of feminism. Unlike earlier waves of feminist protest, the mid-twentieth-century rebirth of feminism inspired enough educated women and their male supporters to successfully challenge many social priorities and institutions. Feminists are sometimes wrongly considered a "special interest" group that is concerned only with the needs and desires of middle-class white women in developed nations. Their efforts, especially since the 1970s, have brought about major changes in legal status, medical treatment, and workplace conditions that benefit all women. Feminism has opened to women professions once available only to men, including medicine, law, the ministry, and academia.

HEALTH-CARE ISSUES

Most ethical issues in patient care and biomedical research affect men and women in the same ways. But some issues affect each differently, due to biological factors, psychological and social occurrences, cultural background, and life experiences. These differences are seen in health-care areas, such as new reproductive technologies; in medical research; and in different treatment options made available to women and men.

General Aspects of the Therapeutic Relationship

paternalistic acting like a father or other authority figure

The relationship between health-care providers and their patients can be described as **paternalistic**. Women have generally been placed below men in the health-care field as well as in other areas of their lives. Behavioral norms and social roles for women have supported stereotypical expectations of greater compliance, dependence, and passivity, especially in relationships with men.

Women's complaints are often dismissed as originating in the mind, implying that they are not to be taken seriously. This may explain some of the gender differences in the way that certain diseases (such as heart disease, which is often not recognized in women) are treated.

Women also often see themselves in ways that are consistent with the view that they are passive or compliant. This parent–child model, in which the patient accepts a passive role in her treatment, can result in reduced communication and less effective care. The dependent, compliant response of the woman patient can interfere with active efforts to recover from trauma or illness. The tension between active participation and the wish to be taken care of is a particular problem for women, because they have been taught to be more passive, and because in most male–female relationships there is a power imbalance.

Specific Problems in the Health Care of Women

Women's bodily characteristics and their reproductive roles lead to many ethical considerations.

Reproductive issues. Decisions about childbearing, the availability of birth control or sterilization, and surgery involving reproductive organs can have important consequences because they affect the family, society, and future generations.

Women generally have not been in a position to set their own reproductive goals, which are often considered to belong to families or to society. Members of families or communities—rather than the affected women—make many of the decisions about medical or surgical procedures that affect women's sexuality and childbearing.

These decisions may also reflect the views or values of the doctor, although they may be presented as medically necessary. A **gynecologist** may assume, without discussing the options with a woman, that she would not want additional children after the age of forty, because she would not want to risk having a child with birth defects. A **hysterectomy** may be recommended even though other options are available. So the choice has been made for her, without her informed consent.

Hysterectomy. Hysterectomy remains the most frequently performed major surgery in the United States, and it is performed more and more on younger women. Gynecologists and others debate the appropriate time for a hysterectomy, often disregarding women's feelings about their uteruses and about losing this particular part of their bodies. The **uterus** has been considered a "useless" and potentially disease-causing organ after the childbearing years. The difficulty of establishing guidelines for when a hysterectomy should be performed is a long-standing problem.

Hysterectomies have been performed on women who have mixed feelings about **contraception** or whose religious or cultural beliefs make birth control a problem. Thus a risky medical procedure may be performed even though safe alternatives are available. Ethical issues raised by this practice include whether a medical procedure should be used for a nonmedical reason, especially with

gynecologist a doctor who specializes in the routine physical care and diseases of the reproductive system of women

hysterectomy surgical removal of the uterus (womb)

see also

Ethics and Law: Information Disclosure, Truth-Telling, and Informed Consent

uterus the organ in women in which the fetus develops; also called the womb

contraception birth control

greater risks, and the problem of whether a woman can make an informed choice in these circumstances.

Contraception. The development of modern birth-control methods has had an important impact on women's lives. Most research on **contraceptives** has been on techniques for women, and the majority of the birth-control devices themselves have been for women. Although this may provide reproductive control to women, who undergo the dangers of pregnancy, women are also exposed to the as-yet-unknown risks and side effects from drugs and other long-term consequences from contraception.

Sterilization. In some situations, laws have permitted the **sterilization** of women considered to be socially undesirable, psychologically deviant, or retarded. This practice has been justified in two ways: (1) the protection of the individual (for example, a mentally retarded woman who is sexually active and may be exploited); and (2) the protection of society against reproduction by "unfit" individuals. Sterilization laws may be a form of discrimination and punishment. In many cases, the conditions considered to be indicators for sterilization, such as psychosis, criminality, and retardation, have not been clearly evaluated or linked to heredity.

Sterilization raises another ethical issue that illustrates the need for informed consent when making decisions that may later be regretted. A woman may not be told that most sterilization procedures, such as having her tubes tied, may not be reversible. But even if she is told, it has been common practice for health professionals to raise questions about a woman's motivation for sterilization, and even to require that she be examined by a psychiatrist. The right of an adult woman to make this choice, when she is appropriately informed, is called into question, and may even result in refusal of the procedure.

Pregnancy. Modern scientific advances have introduced new ethical issues in the care of women during pregnancy and childbirth. These issues have focused on the responsibility of a woman for the health and well-being of her fetus and on the treatment of women without their consent when they are pregnant. In the past, because of the high disease and death rates during pregnancy and childbirth, doctors and parents saw the fetus as a risk to the mother's life and health. As pregnancy and childbirth have become safer, attention has shifted to the health of the fetus, and **prenatal** diagnosis and treatment have become possible. This has caused a shift in viewpoint, so that women who do not agree to prenatal diagnosis and treatment may be seen as not acting in the best interests of their unborn children.

A related issue involves consent for treatment of a fetus before birth, and whether failure to consent can be considered fetal abuse. Viewing the problem in this way implies that the mother and child are separate individuals, and that the mother who refuses treatment has committed a crime against another person—a fetus. A pregnant

contraceptives birth-control devices

sterilization the act or procedure of making a person incapable of sexual reproduction

prenatal concerning the period before birth

woman should not be required to undergo treatment when fetal abnormalities pose no risk to her health, because any treatment carries risks. Doctors should not help one patient by forcing another to take unjustified risks.

Abortion. In the 1973 decision of *Roe* v. *Wade*, the U.S. Supreme Court made abortion (ending a pregnancy) legal in the United States when performed within certain time limits and specific circumstances. The decision then became the responsibility of the pregnant woman and her doctor. Some doctors and hospitals, however, refuse to perform abortions on moral principle or because of limited funds. Although this may not be its intent, the effect of this policy is to discriminate against poor women with little access to transportation, since they may have no alternatives if abortions are not available in their community.

In the abortion issue, the different ethical views may never be resolved. Although some argue that the right to life is absolute from the moment of conception, others disagree and stress that the right to life is the right not to be killed unjustly.

Legal and policy issues focus on a woman's right to privacy, including control of her body, freedom of individual choice, and equality. It is clear that women, because they can become pregnant, face risks and burdens that men do not.

Menstruation. There are still many myths about **menstruation** Whether women have behavioral and mood changes with their menstrual cycles has been a subject of debate for years. Women have been considered unsuitable for important positions on the grounds that they are too strongly affected by changes in their menstrual cycles. Although many women experience no changes during their premenstrual period, others become irritable and have mood swings and other symptoms that disappear when their periods begin.

Premenstrual syndrome (PMS), a nonspecific term for a variety of physical and emotional symptoms, has been claimed to be responsible for various types of social behavior and psychological conditions. Crimes committed, suicide attempts, misbehavior of schoolgirls, psychiatric admissions to emergency rooms, and visits to clinics have been related to the premenstrual period. Most of the early data on PMS are reports by women themselves of their functioning during the menstrual cycle. The data indicate that a small percentage of women feel that their judgment and mental faculties are somewhat impaired during the premenstrual phase of the cycle.

Studies of behavior in relation to menstruation have pointed out problems with much of the research. The bias in choosing to study women with regular cycles, the use of women's own reports instead of observation by researchers, the combination of objective and subjective data, and the problem of determining the hormonal status of the women studied all have brought the validity of the studies into question.

see also

Fertility and Reproduction: Abortion

menstruation the monthly cycle experienced by women of childbearing age

Rape. Women who have been raped are often blamed, disbelieved, and criticized. As a result, they may be reluctant to seek medical care or to report the rape. This often leads to their receiving poor medical treatment, unlike other victims of crime or disaster.

Rape is a crime of violence that is often misinterpreted as a sexual experience. A rape victim may be seriously injured or killed, and her main concern is to survive the rape and protect herself from injury. The lack of consent is crucial to the definition of rape. The victim often has had to demonstrate signs of a struggle or to produce witnesses in order to prove there was no consent—a requirement that does not exist for other violent crimes.

Health-care professionals may have two conflicting roles when dealing with rape victims. Those who first see the victim have to collect evidence so that the rapist may later be prosecuted, yet they

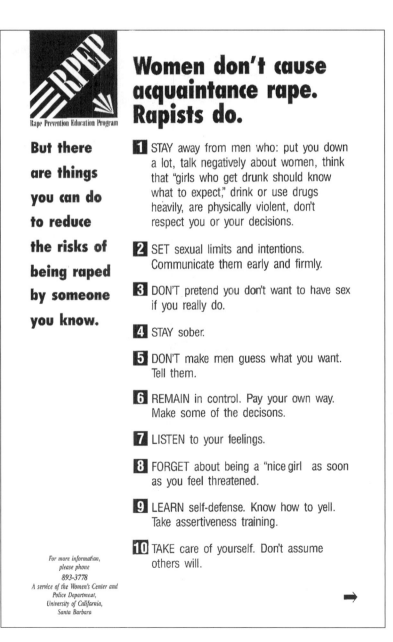

Women don't cause acquaintance rape. Rapists do.

Rape Prevention Education Program

But there are things you can do to reduce the risks of being raped by someone you know.

1 STAY away from men who: put you down a lot, talk negatively about women, think that "girls who get drunk should know what to expect," drink or use drugs heavily, are physically violent, don't respect you or your decisions.

2 SET sexual limits and intentions. Communicate them early and firmly.

3 DON'T pretend you don't want to have sex if you really do.

4 STAY sober.

5 DON'T make men guess what you want. Tell them.

6 REMAIN in control. Pay your own way. Make some of the decisons.

7 LISTEN to your feelings.

8 FORGET about being a "nice girl as soon as you feel threatened.

9 LEARN self-defense. Know how to yell. Take assertiveness training.

10 TAKE care of yourself. Don't assume others will.

For more information, please phone 893-3778 A service of the Women's Center and Police Department, University of California, Santa Barbara

Rape prevention education program poster published by the University of California.

are also obligated to see to the best interests of their patient (the rape victim), who may not want to report the crime. Sensitivity to the patient's condition and wishes may conflict with the need to obtain complete information.

Breast cancer. The increased incidence of breast cancer and the limited resources, until recently, devoted to research and treatment suggest either that it was not taken seriously enough or that it is difficult to get researchers to investigate an issue that does not have special meaning to them.

For some women, the knowledge that they will not necessarily have to have a disfiguring operation has made it easier for them to seek medical attention when they find a lump in their breast. The issue is complicated by the controversy over the safety of silicone implants for women who have had **mastectomies**. It has been contended that evidence concerning breast-implant complications was withheld from those who made decisions about the implants' safety. If this was the case, then patients were deceived into believing they were taking fewer risks than they actually did.

The threat of surgery that may be disfiguring causes many women to delay treatment. The possibility of successful breast reconstruction may help a woman confront and deal with her illness, and may improve her relationships and self-esteem.

mastectomy the surgical removal of a breast

Mental Health

Women have generally been considered more likely than men to have a variety of emotional symptoms and mental illnesses. But the actual rates of mental illness among women and men are difficult to determine because descriptions of symptoms and definitions of mental illness vary greatly. In an early study of how men and women were evaluated and recommended for treatment, researchers reported that doctors' concepts of a mentally healthy, mature man were similar to their concepts of a mentally healthy adult. But their concepts of a mentally healthy, mature woman were more like those for a child. For instance, healthy women were seen as being more submissive, less independent, less adventurous, more sensitive to pain, less aggressive, and less competitive than were healthy men.

Social stereotypes. Although controversy surrounds the methods used in performing this study, and questions exist about whether its results are still valid, there is evidence that social stereotypes still endure and influence patient care. The difference between being labeled "mentally ill" and simply having one's behavior considered to be outside cultural norms or standards is especially important for women, who often feel that any psychological help they receive supports traditional views about women's roles and behavior rather than responding to their individual needs.

Women are typically called on to provide emotional support and assistance for those needing long-term care—children as well as aging parents. Many of these caregivers experience depression, sleeplessness, anger, and emotional exhaustion.

Contrary to popular belief, marriage seems to protect men against mental illness but to make women more vulnerable to it.

schizophrenia a mental disorder characterized by a view of the world that does not match reality, sometimes causing odd behavior

post-traumatic stress disorder a psychological reaction that occurs after a highly stressful event, usually characterized by anxiety and nightmares

postpartum depression a psychological state that affects some mothers after they have given birth

menopause the natural termination of menstruation, occurring usually between the ages of forty-five and fifty–five

Women have been reported to have higher rates of depression, anxiety, and eating and panic disorders, and to exhibit more physical symptoms of mental illness. Men are more likely to be diagnosed with alccholism, substance abuse, **schizophrenia** at a young age, and antisocial personality disorders. There are also gender differences relating to suicide. More women attempt suicide, while more men succeed in killing themselves.

Marital status and mental illness. One difference between men and women involves the effect of marital status on mental illness. Married men and single women have been reported to have the lowest rates of mental illness.

Women experience more physical and sexual abuse than men. This may be an important reason why they suffer higher rates of such disorders as depression, **post-traumatic stress disorder**, and anxiety. Depression has been linked to society's efforts to make women more compliant. The reasons for most gender differences in mental illness, however, are unclear, although it seems that biological, psychological, and social factors all play a part.

The most obvious differences between men and women in rates of mental illness relate to biological and, in particular, reproductive functions. Women, not men, may suffer from reproduction- and menstruation-related disorders such as **postpartum depression**, or symptoms related to **menopause**, hysterectomy, infertility, abortion, and miscarriage.

Women's symptoms and behaviors are often thought to be caused by physical events, including menopause, hysterectomy, or abortion. This demonstrates society's expectations that women will be weak and vulnerable. Such beliefs can unfairly influence researchers' care in gathering and interpreting data.

Conclusion

Social, economic, and technological changes are also changing relationships between men and women and between doctors and patients. Although the paternalistic treatment of women is gradually improving, inequalities still exist in research and clinical care. Some of this results from women's reproductive roles and the possible consequences of allowing women to make independent decisions about their health care.

RESEARCH ISSUES In the early 1990s, people started paying a great deal of attention to biomedical research concerning women's health. They realized that women's-health research had focused almost completely on areas relating to women's reproductive functions. Although this research had been inadequate, research in other areas of women's health had been almost nonexistent. Also, general research findings had been applied to women even though women had not served as subjects in studies.

Protectionism

The policies that have excluded women from research studies have largely resulted from attempts to protect women and fetuses from risks associated with research. There is a long and unfortunate history of the way in which groups of men and women have been abused by research. Because of this history, clear ethical standards, such as those published in the 1978 Belmont Report by the U.S. National Commission for the Protection of Human Subjects of Biomedical and Behavioral Research, have tried to rectify that abuse. This report spelled out the importance of the researcher upholding the ethical principles of respect for individuals, kindness, and justice. It represented opposition to the paternalistic practices of the past. Despite these standards, concerns about women and research are focused on the way in which protective policies have kept women from making their own decisions about their participation in research.

Gender Inequalities in Research

The perception that women's health-care needs have not been properly addressed by the research community led to many of the policy changes regarding women's involvement in research. There are many examples of inadequate attention paid to women in heart-disease studies in the 1960s and 1970s.

Researchers often construct studies concerning women in ways that fail to address women's health concerns fairly. Initial AIDS research focused on the way in which women transmitted HIV to their fetuses or newborns. Scientists paid less attention to how women themselves responded to the virus, even though data show that women with AIDS do not live as long as men do, which may be related to the medical treatment they get.

While data clearly show that women are underrepresented in heart- and AIDS-related research, and there has been little research regarding aging and women, there is less evidence to support the claim that women have not been properly represented in clinical research as a whole.

Perception of Men as the Norm and Women as "Special"

Discussions of research affect how it is viewed and contribute to the creation of biases. Traditionally, society has tended to use the male gender to express universal statements about humanity, thereby creating the idea that men represent the norm for society. As people

see also

General Topics: AIDS

consider women and research, questions arise such as, When do women need to be included in research studies? rather than questions that would reflect gender equality such as, When is it possible to exclude men or women from research?

Though women have gained this "special" status simply by not being men, the fact that they menstruate, may become pregnant, and go through menopause is given as a reason why researchers need to show special considerations if women are to be involved in studies. Alternatively, men could be, but are not, considered special because they do not menstruate, do not become pregnant, or do not experience menopause.

The biases that go along with the way in which language is used (for example, using the word "man" to discuss all of humanity) contribute to the impression that women's involvement in research is often not essential and thus unnecessarily costly and time-consuming.

Procedural Issues

Investigators offer a variety of excuses for women's exclusion from research. Criticisms of the 1989 Physician Health Study's failure to include women in its study were countered with the usual justifications. These involved lower heart-attack rates among women in the age groups studied and the idea that women would not be interested.

For research findings to be meaningful, sample sizes must be sufficiently large and also must represent the group to which the findings will be applied. Any experimental design must use a sample that excludes people whose attributes may interfere with a clear explanation of the differences between an experimental and a control group. The more the subjects in a research sample are similar to each other, the more accurately are the conclusions drawn. However, the more alike the research subjects are, the less easily the findings of the study can be generalized to include society at large.

Consequently, the popular view that women are special has caused them to be excluded from research studies—because to include studying women's unique factors would demand larger and more expensive sample sizes. In such cases, making research easier has been deemed to be more important than understanding women's concerns (for example, how women's hormonal differences affect responses to drugs or other medical treatment). As a result, women's health care is frequently based on findings from male subjects, whose hormonal makeup is different.

Conclusion

Between 1990 and 1994, there were major policy and legal changes involving research on women. The National Institutes of Health (NIH) Revitalization Act of 1993 declared that women "must be included

There Is No Shallow End.

Parkland
Do More. Be More.

▲ Advertisement for women's health.

in all NIH-supported biomedical and behavioral research projects unless a clear and compelling rationale and justification is established." This statement represents a major shift from previous policies and perspectives on women as research subjects.

These guidelines must be interpreted so that a balance can be achieved between developing diverse enough samples for study findings to be applied to as many people as possible while maintaining workable research methods. More than women's representation is necessary for data to be applicable to women. The NIH guidelines demand that minorities also be represented. Attention also needs to be given to the importance of research subjects' ages.

The early 1990s represented an awakening to the importance of gender in research. However, sticking to the guidelines for including women in studies must be coupled with reasoned judgment and creative question-asking. A broad, careful view is necessary as research continues to aid in improving the health of women and society

Related Literature

W. H. Auden's poem "Miss Gee" (1968) describes a childless woman who always seems to be buttoned up and self-protective. Finally, an internal pain drives her to see a doctor who diagnoses cancer. He believes cancer strikes childless women and men when they retire, as if somehow their lives have lost meaning and made them vulnerable to attack.

◆◆◆

Toni Cade Bambara's short story "A Girl's Story" (1974) describes what happens when a girl, Rae Ann, gets her first menstrual period and does not have any idea what is happening to her. Her mother has died long ago. Neither her grandmother nor anyone else has prepared Rae Ann for this experience, so she is afraid she is dying. To make matters worse, when her grandmother discovers her bleeding in the bathroom, she accuses her of having an abortion. It turns out that a botched abortion was the cause of Rae Ann's mother's death, and the grandmother jumps to the wrong conclusion, not helping the girl at all with her confusion and fear.

◆◆◆

In "The Space Crone" (1989), Ursula Le Guin states that women after menopause should see themselves as gaining a third identity, having grown past virginity and past motherhood. Now a woman moving through menopause must become "pregnant with herself"—growing into wisdom, as one who has experienced the whole human condition.

Stages of Life

INFANTS

This entry consists of four articles explaining various aspects of this topic:

neonatal concerning the first month after birth

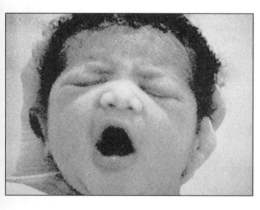

prenatal concerning the period before birth

MEDICAL ASPECTS AND ISSUES IN THE CARE OF INFANTS

The most often discussed ethical problems having to do with infants involve decisions to withhold or stop life-sustaining medical treatment. Such discussions usually focus on the relatively small world of the acute-care hospital or **neonatal** intensive-care unit (NICU). These cases are talked about in terms of the morality of case-by-case decisions. Such life-and-death decisions are controversial because, in the United States and some other countries, the patient's expressed wish has been the popular legal and moral standard.

General Ethical Issues

Infants cannot express their own preferences or participate in such decisions, so these must be made by others. Commonly, parents were given the power to decide, but modern legal developments provide children with their own rights. Parents are not permitted to make decisions that would directly endanger the life, or seriously risk the health, of their child. In other words, they must make decisions that are in the child's best interest. But determining what that interest is may not be clear, and different people may judge it differently in different situations.

Because of this emphasis on individual patient decisions, less attention is paid to broader issues that also influence child health, such as the fair distribution of medical resources, or the limits of society's responsibility to improve the health of all infants. In the United States, the overall infant death rate is nine per 1,000, yet African-American infants die at nearly twice that rate. The imbalance between infant death rates for the most and least fortunate Americans is due in part to an inequitable health-care system. Preventive care and **prenatal** care are unavailable for many women, yet access to neonatal intensive care is guaranteed by law.

Discussions of broad issues usually involve terms that are hard to apply to individual situations, while discussions of individual

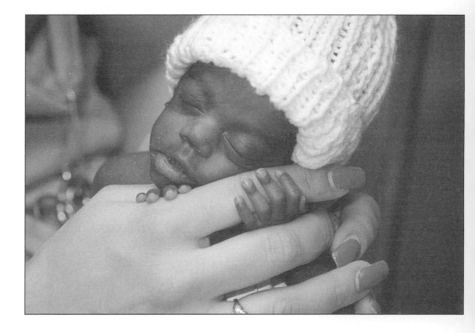

▶ A premature, low-birth-weight baby.

situations often ignore larger issues of medical-resource distribution and social justice.

Definitions

Infant mortality is defined as death before one year of age. In some ways, the infant death rate defines what neonatal medicine can achieve. At the lower limits, infant death rates give an idea of what is "natural"—that is, how many infant deaths cannot be prevented by medical or social interventions. Short of this limit, the infant death rate suggests something about how many infant deaths a society is willing to endure.

Gauging infant mortality. The infant death rate is an unusual statistic because it measures an effect (death) that has various causes (genetic diseases, illnesses, injuries, and so on). Comparisons of infant death rates between countries or ethnic groups allow comparisons of the adequacy of social services. The infant death rate can also be used to check how cost-effective particular medical treatments are. High infant death rates have been used in the United States to argue for a wide variety of interventions, including global access to prenatal care, widespread use of neonatal intensive care, increased funding for Head Start programs; national programs to provide nutritional supplements to women, infants, and children; family-planning programs; legalized abortions; and decreased military spending.

Infant mortality is divided into two types, neonatal mortality and postneonatal mortality. Neonatal mortality is death before twenty-eight days of age. Postneonatal mortality is death between twenty-eight days of age and a baby's first birthday. Factors lowering neonatal death rates can be divided into those that increase

> The infant death rate is often used as a way of measuring social welfare.

infectious spread by a virus or other microorganism

premature born after a pregnancy of fewer than thirty-seven weeks

congenital existing at or dating from birth

postneonatal concerning the period following the first month of life

It's enough to make any parent lose sleep.

SUDDEN INFANT DEATH SYNDROME, or SIDS, claims the lives of nearly 6,000 infants in the United States every year. In fact, SIDS is the leading cause of death in infants one month to one year old.

But you can help reduce the risk of your baby's dying from SIDS simply by placing it on its back or side to sleep.

Talk to your doctor about SIDS and how you can help reduce the risk. For more information, write Back to Sleep, P.O. Box 29111, Washington, D.C. 20040, or call, toll-free:

1-800-505-CRIB
National Heart, Lung, and Blood Institute
National Institute of Child Health and Human Development
U.S. Public Health Service
American Academy of Pediatrics   SIDS Alliance   Association of SIDS Program Professionals

A poster about SIDS, a condition that kills six thousand infants in the United States every year.

parenteral occurring outside the intestine

birth weight and those that decrease mortality for babies born at low weights. Postneonatal mortality is related less to birth weight than to immunization rates and the presence of **infectious** disease.

Higher birth weight is strongly associated with decreased infant mortality because babies of low birth weight (LBW) are generally either **premature** or ill. Prematurity is associated with death because crucial body functions or body organ systems may be underdeveloped and not function properly at birth.

The impact of social services on infant mortality. Availability of family-planning services (programs that provide birth control and abortion) has been linked with decreases in infant mortality. Family-planning programs allow women to decide when to become pregnant and have children, leading to improved maternal health, babies of higher birth weight, and lower infant mortality.

Laws, programs, or judgments made by the government, policymakers, or society as a whole affect trends in birth weight. In the United States, widespread neglect of preventive medicine and family planning has led to an exceptionally high need for neonatal intensive care. As a result, doctors in the United States have developed treatments that decrease birth-weight-specific mortality.

Postneonatal infant mortality is caused largely by a few diseases. The most common causes are Sudden Infant Death Syndrome (SIDS) and **congenital** abnormalities. Deaths due to congenital abnormalities are generally not preventable. Deaths due to SIDS and other diseases are considered preventable.

Postneonatal death rates fell steadily throughout the twentieth century. The U.S. postneonatal death rate declined from over sixty postneonatal deaths per 1,000 live births in 1900 to fewer than four deaths per 1,000 live births in 1987.

There will always be sick newborns, even in countries with extensive social programs and universal access to family planning and prenatal care. Worldwide funding decisions cannot eliminate individual dilemmas, but they can affect their frequency and the context within which they must be resolved.

Specific Ethical Issues

In developed countries, low birth weight (LBW) is the most important factor associated with infant mortality. LBW, low birth weight is defined as less than 2,500 grams (or about five pounds and eight ounces) at birth. Very low birth weight (VLBW) is birth weight below 1,500 grams (or about three pounds).

The treatment of LBW. Major advances in the treatment of LBW infants occurred during the 1970s, 1980s, and 1990s. Two medical interventions in particular have made the survival of VLBW babies possible: mechanical ventilation and total **parenteral** nutrition. Mechanical ventilation allows treatment of respiratory distress syndrome (a common problem in premature babies) by pumping

intravenously entering by way of a vein

prognosis a prediction about the future course of a disease and the patient's chances of recovery

viable able to survive outside the mother's womb without artificial support

oxygen-rich air into the infant's lungs. Parenteral nutrition enables infants who are too underdeveloped to eat to get ample nourishment **intravenously** and to survive and thrive.

Most babies who die in NICUs do so in the first seventy-two hours of life, even if they get all available treatment. Almost any group of infants with a high likelihood of dying will include babies who survive in spite of a discouraging **prognosis**. As a result, many caregivers choose to provide intensive care to all **viable** babies and allow the babies to "declare themselves," either by stabilizing or by deteriorating. Treatment issues for newborns with two common conditions—respiratory distress syndrome and intraventricular hemorrhage—demonstrate the ethical factors involved in the care of infants.

Respiratory distress syndrome (RDS). Respiratory distress syndrome (RDS) is the most common condition affecting premature babies. It is caused by lungs that have not yet fully developed to breathe on their own. Treatment of RDS requires high-pressure mechanical ventilators using high concentrations of oxygen. This treatment can damage the infant's delicate lung tissue, leading to the lung disease bronchopulmonary dysplasia (BPD).

BPD illustrates the difficult ethical decision making in neonatal intensive care. BPD can lead to early death, prolonged disease with eventual death, survival with permanent lung disease, or complete recovery. Prognosis for any individual newborn is uncertain at birth. Although predictions become more accurate as children get older, uncertainty remains. So decision makers must consider not only the

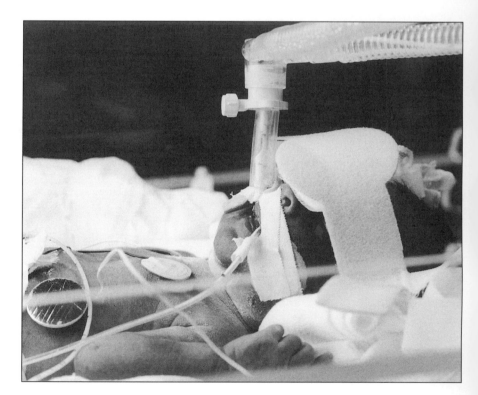

▶ An infant lies in an intensive-care unit.

severity of illness and its impact on quality of life but also the chances of survival or death. In evaluating quality of life, the availability of home care using life-sustaining technology offers some new hope for these children, although the hope comes at an enormous economic and emotional cost.

Intraventricular hemorrhage (IVH). Like RDS, intraventricular hemorrhage (IVH) is not so much a disease as a condition that arises because of the immaturity of the circulatory system in the fetal brain. The outcome of infants with IVH depends on the degree of brain injury associated with the hemorrhage. The more immature the infant, the more likely it is that a hemorrhage will occur.

Ethical issues in newborns with RDS and IVH. Neonatal RDS and IVH both present ethical dilemmas at two points. First, when premature infants develop RDS or IVH, questions arise about whether life-sustaining treatment should be stopped. Doctors may decide not to begin mechanical ventilation for babies weighing less than 800 grams (one and a half pounds). These decisions may be based on concerns about the prognosis for survival, about the quality of life for survivors, about the fair distribution of limited intensive-care resources, or about the economic and emotional cost to families. Often, doctors respond to any uncertainty with a wait-and-see attitude, continuing treatment until it is possible to make a better prediction. This leads to the second ethical dilemma.

If decisions to withhold treatment are put off early in the course of treatment because of uncertainty, doctors must define an outcome that would be considered unacceptable at some point during an illness. For infants who develop BPD, a number of long-term results may be so considered. Such infants have weak lungs and an increased frequency of lung infections and asthma. Some remain dependent on a ventilator. Children with BPD grow slowly, and BPD may be associated with other developmental delays. Any one of these factors may justify a conclusion that prolongation of life is not in the child's best interests. For infants with IVH, decisions reflect the degree of nerve damage and the presence of associated nervous-system problems, such as seizures.

hemorrhage heavy and uncontrollable bleeding

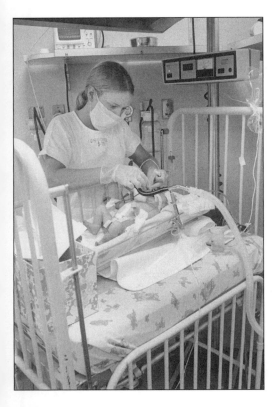

△ A nurse holds a tube to the face of a newborn baby in need of respiratory assistance.

Conclusion

Any discussion of the ethical dilemmas in the care of infants must consider both general and specific ethical issues. The general issues are harder to frame, since they involve society's unspoken choices concerning the value of infant lives. Nevertheless, the broader issues can be successfully examined by comparing events and policies in different countries or societies as if those were willful decisions made by ethical individuals.

In evaluating specific ethical issues, it is easy to miss the forest for the trees. Any medical case must be handled with the understanding that no outcome is certain. Those who try to find answers for these questions need to base arguments not only on moral principles but also on statistical realities. Individual circumstances differ widely, and the highly singular knowledge needed to evaluate any baby's prognosis can be intimidating.

HISTORY OF INFANTICIDE In many societies, infanticide—the killing of infants—was not only allowed but also sometimes promoted as a solution to the problem of unwanted babies, both sick or healthy.

Infanticide in Ancient Times

In ancient Greece, people measured an infant's worth by its potential usefulness to society. The philosopher Plato wrote that deformed newborns should be "hidden away, in some appropriate manner that must be kept secret."

It is difficult to distinguish between infanticide and abandonment, which may or may not have intended death. Historical evidence is not clear as to whether abandoned infants usually died or if those who abandoned them meant for them to die. It is likely that direct infanticide was practiced for both **eugenic** purposes and population control. Laws neither forbade the killing of deformed infants nor protected healthy infants from death by exposure.

Evidence suggests that infanticide was practiced without punishment in Roman society. While Romans continued to dispose of deformed infants for eugenic and economic reasons, another motivation stemmed from the Roman belief in unnatural events. The Greeks saw deformities in newborns as natural occurrences. In contrast, the Romans viewed them as threatening "signs" that needed to be destroyed in order to rid the community of guilt and fear. Roman society seemingly accepted infanticide as a reasonable disposal of deformed infants for superstitious motives.

Early Jewish and Christian Traditions

Jewish scholars were among the first to clearly condemn the killing of infants. Jews believed that Yahweh (God) created humans in his own image. All human life was sacred from the moment of birth. The Torah (first five books of the Hebrew Bible) speaks of deformed individuals as Yahweh's creations, and it orders protection of the blind, the deaf, the weak, and the needy. Human life had built-in value through divine endowment—not merely value by virtue of social use, as in classical Greek and Roman society.

Christianity significantly changed public attitudes toward infanticide. Christians inherited the Jewish principle that humans were

eugenic having to do with selective breeding of humans

Genetics: Eugenics

Before Christianity came to England, it was an accepted practice for a father to "dispose" of an unwanted or deformed child. The decision was solely the father's, even though the task may have been carried out by other individuals in the household.

original sin the state of sin (or lack of grace) that characterizes all human beings as a result of Adam's disobedience

divinely created, including the emphasis on the sacredness of all human life. Two other Christian concepts important for their effect on the practice of infanticide were **original sin** and its related ritual of infant baptism, which probably became common in the third century. Christians believed that unbaptized infants who died were condemned to eternal hell. [Baptism at birth or soon after protected infants because it bestowed on them a high worth and equality in the community of Christians. So, baptism served as an important deterrent to both abandonment and infanticide.]

Medieval Period

Christianity's beliefs mixed with myth, superstition, and folklore during Europe's medieval period (roughly the 8th through 14th centuries). This blending had serious implications for deformed infants and the practice of infanticide. Some people thought that the parents' sexual behavior generated abnormal births. Others believed that sexual relations during menstruation, pregnancy, or when a woman was nursing a baby resulted in serious consequences for the unborn. In addition, the birth of an abnormal infant was sometimes attributed to the devil's interference. People saw such births as the product either of a sexual relationship between the mother/witch and the devil or of baby-switching by the devil as punishment of the parents' sins. Society held parents, particularly mothers, as morally responsible for their infants' deformities.

▲ Mary breastfeeding the baby Jesus in early medieval art.

In medieval Europe, laws against infanticide usually considered the crime to be murder. But suffocating an infant sleeping in the parents' bed—the most frequent cause of infanticide—was easy to hide and intent was nearly impossible to establish. This made prosecution very difficult.

Throughout the medieval period, infanticide was regulated largely by church courts rather than civil courts. Penalties imposed by the Church on married women convicted of infanticide were often light. Punishment involved penance and was comparable to that imposed for such offenses as premarital and extramarital sex.

illegitimacy the state of being born of unmarried parents

foundling home a home set up to accept abandoned infants

Although direct infanticide was practiced to some extent, the more common cause of infant death during the Middle Ages was abandonment. The distinction between infanticide and abandonment became increasingly more important because abandonment was generally regarded as a less serious offense, punishable only if the child died. In the early Middle Ages, abandonment was widespread, mostly due to poverty and **illegitimacy**. Although some church leaders believed it was equal to infanticide, two forms of abandonment were quite common: donating infants to the Church and leaving infants at **foundling homes**. From a Christian point of view, both customs were improvements over the morally objectionable practices of exposure and infanticide.

Cities and towns established foundling homes to lessen the practice of exposure and to provide a humane solution to infanticide. In reality, however, the foundling home often was equal to a death sentence by neglect and disease. Once infants arrived at a foundling home, they frequently were sent to the country with a woman able to nurse the baby. Often, this woman was negligent and more interested in a steady flow of babies, for which she was paid, than in providing care. Death rates were high, especially for female infants.

Renaissance and Reformation

During the sixteenth and seventeenth centuries, there was a unified effort throughout Europe to stop infanticide. The dramatic rise in reported cases may have been because city living made it more difficult to kill infants secretly. Authorities announced harsh laws aimed at ending the practice and increased the prosecution of murdering mothers.

In 1532, Emperor Charles V of the Holy Roman Empire issued a law known as the Carolina. This was the first attempt to strengthen and unify infanticide laws. The law decreed that those found guilty of killing an infant were to be buried alive, lanced, or drowned. The law also made hiding a pregnancy a crime, as it was assumed that such secrecy indicated murderous intentions.

Puritan interests played a role in the increased eagerness to punish illegitimacy. The 1623 English infanticide law, influenced by the Puritan members of Parliament, allowed courts to convict on the basis of indirect evidence of concealment and earlier sexual misconduct. The law assumed that the child was born alive and then killed unless the mother could prove otherwise. The majority of infanticide victims during this period were born outside of marriage.

Eighteenth and Nineteenth Centuries

In the eighteenth century, a steep drop occurred in formal charges of infanticide. The courts during this time showed greater mercy toward those accused of killing their children. Attitudes toward parenting changed as well, with a new emphasis on emotional nurturing. Having poor women nurse wealthy women's babies became less frequent, while keeping infants with their mothers became more common. At the same time, the growing needs for workers generated by the Industrial Revolution resulted in larger families among the poor and working classes.

"Benefit of Linen." Lawyers developed new defenses for the mother suspected of infanticide. One of the first of these defenses, known as "benefit of linen," held that if the mother had made linen for the baby before its birth, then she had no plan to kill it. This

A nineteenth-century illustration, titled "Disposing of the body," shows a woman carrying the body of an infant out of town and into the countryside. Her worried and furtive demeanor are meant to indicate that she is looking to get rid of the baby's body.

Although available statistics do not show an increase in the incidence of infanticide during the Victorian era, newspapers, professional journals, and parliamentary papers of the time all reflect growing concern of both government and the public. Books and dailies all described infanticide as a "growing epidemic" and as a "national disgrace" for England.

The anger over infanticide was due, in part, to the fact that the crime frequently went unpunished. Juries were very reluctant to bring in murder convictions against women accused of infanticide. Juries often chose to overlook even the most shocking evidence of guilt, either convicting the defendant on the minor charge of "concealment of death" or aquitting her outright. The law enabled such decisions, because the standard of evidence to prove infanticide was almost impossible to meet.

euthanasia *mercy killing*

Death and Dying: Euthanasia and Sustaining Life

line of argument became very popular after 1700 and virtually guaranteed the dismissal of charges. Another major defense commonly used was the "want of help" plea. Various accidents, such as failure to tie the umbilical cord, falls of either the mother or baby, illness of the mother, and unanswered cries for help, all effectively helped to sway jurors.

Despite successful reform efforts, infanticide did not disappear. During the nineteenth century, high rates of illegitimate births continued, as did infant killing. Dead infants found in outhouses, parks, rivers, and other public places fueled the perception that infanticide was reaching unacceptable proportions. This perception may or may not have represented an actual increase in the incidence of the crime, but it did serve to provoke a huge public outcry.

Many doctors led reform efforts. In his 1862 essay on infanticide, William Burke Ryan wrote passionately against the horrors of infant murder. He and a few other doctors formed the Infant Life Protection Society.

Twentieth Century

The most notorious instances of infanticide in the twentieth century were committed secretly in Nazi Germany, under the power of the Committee for the Scientific Treatment of Severe, Genetically Determined Illness. The Nazis required doctors, nurses, and teachers to register all children with congenital deformities or mental retardation. Failure to comply meant civil penalties or imprisonment. Defective children were removed from their homes and routinely put to death at hospitals by morphine injection, gas, lethal poisons, or starvation. To ensure secrecy, workers immediately cremated the bodies.

In the 1920s and 1930s in the United States there were proposals for legalized **euthanasia**, which was justified primarily as a way of limiting the social costs associated with defective infants. But as the realities of the Nazi extermination programs began to surface in the United States in the 1940s, promotion of euthanasia in general began to decline.

ETHICAL ISSUES The birth of a baby can be one of the most satisfying, fulfilling experiences of a parent's life or a couple's marriage. Unfortunately, sometimes the months of dreams and plans for a normal baby turn out to be false hope. Even when prenatal diagnosis has already suggested that the baby will not be normal, there may still be surprise and disappointment at the range of medical problems and the degree of nervous-system damage the child has. When the parents had no opportunity to prepare, the birth of a premature or disabled infant can have a huge emotional impact that severely tests parents' beliefs, values, and future hopes.

In the early 1960s, neonatal intensive care units (NICUs) were not yet very advanced. In 1963, Jacqueline Kennedy gave birth to a son, Patrick Bouvier Kennedy, five weeks before the due date. The baby weighed 2,100 grams (4½ pounds) and died the next day despite all efforts to save him. Today, an NICU would have no trouble saving such a baby.

The birth of such a baby can also point up different ethical viewpoints among parents, doctors, and others regarding the value of infants with life-threatening medical conditions. This is especially true when the projected future lives of these children are filled with a mixture of mental and physical disabilities. For many people, such cases raise important questions: What is the moral status of infants with mental and physical disabilities? Should all of these infants receive life-sustaining medical treatment regardless of the seriousness of their medical conditions? Is there any moral difference between withholding and stopping life support? Would it be justifiable to kill any of these infants?

The Modern Period

The birth of a healthy, normal baby is a cause for celebration because the baby offers future promise for the family. Before modern medicine, a successful birth meant that the mother had survived the dangers associated with pregnancy and childbirth, dangers that posed a serious risk to the mother's health and life in every pregnancy.

However, not all births are celebrated. In many societies and in almost all historical periods, very young infants, female infants, illegitimate babies, and infants and older children believed to be "defective" in some way are still killed (see previous section). Parents have historically had several possible reasons for killing one or more of their children. Some of them have killed for economic reasons. A dead child would mean one less mouth to feed. Others have killed their infants because of social customs and pressures. One might kill an illegitimate child, an "extra" child beyond a certain number, or another female child.

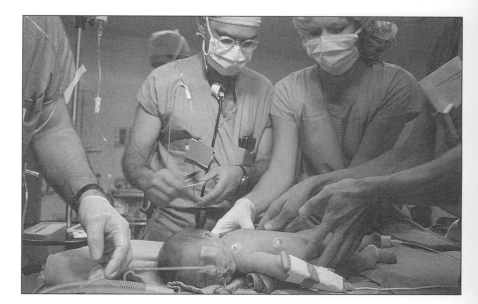

▶ Taking advantage of advanced medical therapies, doctors remove a tumor on a newborn baby's spine.

Two features of traditional child-killing practices remain a part of the modern world. First, infants are still sometimes killed by their parents or abandoned without food, shelter, or protection. Second, even for parents who cannot imagine killing their own children, the birth of an extremely premature or severely disabled infant is a mixed blessing. For that reason, parental decisions about medically prolonging a child's life frequently involve concerns about the nature of the family as well as considerations about the welfare of the child.

In many parts of the world, such decisions, whether made by a child's parents or its doctors, are quite similar to decisions made about sick and disabled children in earlier times because many countries still lack the medicines, the medical personnel, and the technology that are common to the rest of the world. In technologically developed countries, by contrast, the development of neonatal intensive-care units (NICUs), **pediatricians**, sophisticated medical technology, new medicines, and new surgical techniques have brought new opportunities and challenges to doctors, parents, nurses, and all others interested in prolonging the lives and improving the health of critically ill children.

pediatrician a doctor who specializes in the development, care, and diseases of children

Attempts at Regulation in the United States

The decision not to use medical technology to sustain an extremely premature or severely disabled infant's life is usually difficult and sometimes controversial. One effort at regulating selective decisions not to treat took the form of two sets of published federal rules during the administration of President Ronald Reagan. The 1983 "Baby Doe" regulations and the subsequent child-abuse regulations (1985) differed in legal philosophy, practice, and influence. Yet both agreed on the ethical perspective that should govern life-and-death decisions made in NICUs and pediatric intensive care units (PICUs). That perspective was that every infant should be given life-sustaining treatment, no matter how small, young, or disabled the infant might be. The only exceptions would be infants who were permanently unconscious, fatally ill, or able to be saved only with extreme, inhumane treatment.

A second effort at regulation was made by the U.S. President's Commission for the Study of Ethical Problems in Medicine (1983), the American Academy of Pediatrics, and several writers on ethics in pediatric medicine. Given the difficulty of some pediatric cases and the life-and-death nature of selective nontreatment decisions, the common recommendation was to have an ethics committee consult on the cases and give advice to the doctors involved. In truly difficult cases, the most cautious procedure for decision making is to reach agreement by a diverse committee that is knowledgeable, impartial, emotionally stable, and consistent from case to case.

In 1982 Baby Doe was born with Down syndrome and an obstruction in the esophagus. The parents refused the surgery that would have corrected this defect, and the baby died after six days. Immediately after the baby's death, President Ronald Reagan instructed the U.S. Department of Health and Human Services to issue a ruling that would protect handicapped babies from discrimination in treatment in the future. This ruling became known as the Baby Doe Regulations (1983).

Baby K

In October 1992 Baby K was born with a congenital defect known as anencephaly. This is a condition where the infant has a skull but essentially no brain in it, except for the primitive brain stem, which controls breathing, blood pressure, and temperature. Baby K could not breathe on her own at birth. Usually, these infants die with a few days or weeks.

Baby K's mother wanted everything done to extend the life of her baby. The hospital went to court to have the life support removed, but the court ruled that Baby K should not be discriminated against. With the help of a breathing machine, Baby K lived a few months and then died. This effort to extend her life cost more than $500,000.

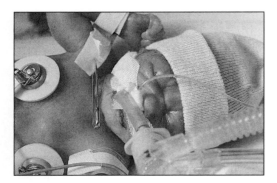

▲ A premature infant is attached to a battery of monitors in the neonatal intensive-care unit of a hospital.

Some themes and problems are common, as decision makers in technologically advanced countries face the difficult choices presented by premature and disabled infants. First, the ongoing technological development of pediatrics has resulted in improved death and disease rates for many infants and young children. Second, new surgical techniques have resulted in the extension of life for many infants who would have died without surgery only a few years ago. Third, these achievements have started a trend in some pediatric specialties toward overtreatment of premature and disabled infants. (This trend seems to be contrary to the best interests of some of these children.) Fourth, doctors and parents still frequently face an inescapable problem: medical uncertainty regarding the degree and range of disability a child will have if it survives with treatment.

Philosophical Perspectives on the Moral Status of Infants

Ethical perspectives on infant care are strongly influenced by views regarding the "personhood" status and moral standing of infants, whether premature, disabled, or normal. What kind of being is it whose life, health status, or death is at stake in the decisions made by doctors and parents? Does a newborn count as a person, in the same way that an older child and an adult count as persons? Or are questions about personhood irrelevant in terms of the moral standing that adults choose to grant infants? In terms of moral standing, what kinds of moral rights do infants possess? Do infants have full moral standing, making them morally equal to adults? Is the moral standing of newborns to be understood as somehow less than that of adults but more than that of fetuses? Or are fetuses, newborns, and adults to be thought of as morally equal?

For many modern philosophers, questions related to the moral standing of infants can be answered only after "personhood" has

see also

Fertility and Reproduction: Fetus

been defined. One approach is to define "person" as meaning "a living being with full moral standing." According to this definition, all persons have such standing, leaving open the question of just which characteristics give this standing.

What Makes a Newborn a Person

Three positions link the personhood status of newborns with the moral standard granted to infants.

All newborns are persons. The first position holds that all newborns, whether normal or impaired, count as persons in the same way that you and I do. According to this view, the personhood of newborns is merely an extension of the personhood possessed earlier by fetuses. Newborns, like all other persons, have the moral right not to be killed or prematurely allowed to die, since the possession of personhood includes full moral standing, regardless of age.

Moral characteristics define personhood. The second position holds that in order to count as persons, infants must possess the inherent qualities of consciousness, self-awareness, and rationality. If infants lack these basic traits, they have a personhood status that is closer to that of human fetuses than to that of older children or adults. Holders of this view claim that all newborns, including normal babies, fail the neurologic tests for personhood and so are to be classified as nonpersons. In this view, all newborns lack the **cognitive** qualities that make a human into a person. In addition, the idea of potential personhood is discarded, largely because of the argument that personhood cannot be possessed in varying degrees. Holders of this second view also claim that only those who have the characteristics of persons possess the rights of persons, including the right not to be killed or prematurely allowed to die.

Newborns as potential persons. The third position stands between the other two. It identifies the same characteristics of personhood, but according to this view, most newborns are to be regarded as potential persons, not yet possessing the status of actual persons but on the way to possessing basic traits of personhood through normal human development. Agreeing with supporters of the first two positions on the connection between personhood status and moral standing, philosophers holding the third position maintain that when infants subsequently become persons, they will acquire full moral standing. Until that time, including during the neonatal period, they are regarded as having a right not to be killed, allowed to die, or significantly harmed, precisely because they will subsequently and naturally become actual persons.

The implications of such definitions. The differences in these philosophical views have practical consequences for the ways in which adults value the lives of infants, including infants who may be extremely premature or severely disabled. Supporters of

cognitive involving the proces of knowing or of acquiring knowledge; pertaining to the state of awareness, perception, and reasoning

the first position tend to call for life-sustaining treatment to be administered to all infants in NICUs regardless of birth weight or health status. They believe all infants are actual persons in possession of the full array of moral rights common to persons. By contrast, any parents or doctors in NICUs who regard newborns as nonpersons are likely to be ready to withhold or stop treatment much more quickly, because the lost infant lives do not yet count for much morally. For supporters of the third position, the concept of potential personhood provides an intellectual framework in which difficult judgments make some sense. In this view, at least part of the difficulty in making decisions to provide or to stop life-sustaining treatment has to do with judgments about whether a particular baby has the potential to become a person in the normal course of development.

Social Perspectives on Personhood

Other perspectives on the moral status of infants, some of which are based on religious ethics, suggest that the philosophical debate about the personhood of infants is restrictive and of little practical worth. One common view is that the moral standing of infants cannot depend on whether they meet a philosophically strict definition of personhood, because all infants fail to meet that standard. Rather, what is important is a social understanding of "person" according to which infants qualify as viewed by their parents, doctors, and others.

Another widely held view is that the personhood question simply does not apply to infants. Rather, what is important is that infants are understood to have moral standing as "fellow human beings." Supporters of this view may see fetuses and infants as having equal moral standing as human beings. Or they may have a developmental view in which viable fetuses and infants, but not nonviable fetuses, have equal moral standing as human beings. Either way, infants are regarded as having the same kinds of moral rights that older human beings have, including the right not to be killed or allowed to die prematurely.

Perspectives on Stopping Life-Sustaining Treatment

The ethical perspective in the "Baby Doe" and child-abuse regulations was only one of such perspectives on the medical care of infants that have received a lot of attention. Other ethical perspectives have also been widely held, both before and after the federal rules became policy.

Quality of life issues. For some people the important ethical question is not whether a given infant can be saved through medical treatment. Rather, the important question is what quality of life the child will probably have later. The question is especially

Baby Doe regulations have made sure that newborn infants with severe disabilities are treated like any other baby. This has resulted in treatment of hopeless cases by doctors who are afraid they may otherwise be sued. Sometimes the doctors do not even tell the parents what they have done until after the treatment.

In 1998, a Texas court awarded the parents of a severely disabled girl $70 million because a hospital had insisted on treating their child even though she weighed only 1½ pounds at birth.

meaningful if the child's future is predicted to be dominated by severe impairments, multiple surgeries, and many other medical problems.

The child's best interest. A closely related ethical perspective focuses on a child's best interests. For those holding this position, the important question is whether the life-sustaining treatment that could be given to endangered newborns will, on balance, provide the infants with more benefits than burdens. This position's strength is that it focuses on the patient's best interests, not the interests of the family or society.

Ethics in Nontreatment Decisions

Ethical perspectives play a major role in selective nontreatment decisions. The dominant perspective in individual cases varies from hospital to hospital, doctor to doctor, parent to parent, case to case.

Treatment for all living infants. The first perspective calls for life-sustaining treatment to be administered to all infants who are conscious, not dying, and for whom treatment is not "virtually futile and inhumane." The federal regulations that reflect this perspective have been largely unenforced throughout the United States.

The reasons for its continuing influence are twofold. First, this perspective is consistent with the reasons that motivate doctors who work with newborns to do the work they do: to prolong and enhance the lives of the youngest, smallest, most disabled, and most vulnerable human beings. Second, this perspective offers the simplest way of dealing with the many problems that make up the NICU. It minimizes medical and moral uncertainty in cases, the role of parents as decision makers, and any considerations of the harm that may be done through prolonged, aggressive efforts to save endangered young lives.

Parents as decision makers. The second influential perspective stresses the role of parents as decision makers. Supporters of this view rarely suggest that parents alone should make the selective nontreatment decisions that could cause the deaths of their children, or that parents should be given unlimited discretion in making such decisions. Rather, the claim often is that parents should, in response to appropriate medical information and advice, have reasonable discretion in making a life-and-death decision regarding their child in the NICU. They are the ones, after all, who may be responsible for the huge financial costs of intensive care. They are the ones, in addition to the child, who will have to deal with ongoing medical problems, repeated hospitalizations and surgeries, abnormalities, and developmental delays.

The child's best interests. The third perspective is the patient's-best-interests position. Supporters of this position acknowledge the medical and moral uncertainty inherent in many cases, affirm an important role for parents as decision makers, and

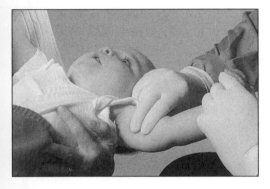

▲ An emergency medical technician checks the pulse of an infant.

recognize that the same medical and surgical treatments that produce great benefit for some patients can produce harm for others. In contrast to the parental perspective, proponents of this view stress that the decision-making focus in neonatal and pediatric cases should be the best interests of the patient, even when they conflict with those of the parents.

The Emerging Mainstream Perspective

If any of these positions can be correctly deemed the mainstream ethical position, at least in the United States, it is the patient's-best-interests position. Supporters of this position are concerned about the treatment-related harms that sometimes occur when doctors persist, perhaps under the influence of the federal regulations, in overtreating infants who have extremely low birth weights and severe disabling conditions but who are neither unconscious nor dying. At the same time, proponents of the best-interests view are reluctant to grant the parents of premature and disabled infants as much discretion in deciding to stop life-sustaining treatment as some parents would like.

The criteria for decision making. The best-interests position relies on eight variables that help to determine whether to start, continue, or stop life-sustaining treatment:

1. the severity of the patient's medical condition, as determined by evaluation and comparison with (a) all infants and (b) infants having the same medical condition
2. the achievability of treatment, in an effort to determine what is meant by "beneficial" treatment in a given case
3. the important medical goals in the case, such as the continuation of life, the effective relief of pain and suffering, and the improvement of disabling conditions
4. the presence of serious brain impairments, such as permanent unconsciousness or severe mental retardation
5. the extent of the infant's suffering, as determined by the signs of suffering that infants show by means of raised blood pressure, increased heart rate, degree of agitation, and crying
6. the many other serious medical problems
7. the life expectancy of the infant, because some of the severe congenital problems involve a life expectancy of only a few weeks or months; and
8. the ratio of treatment-related benefits and burdens to the infant, a medical and ethical "bottom line" for determining whether life-sustaining treatment or the stopping of such treatment is in a particular infant's best interests.

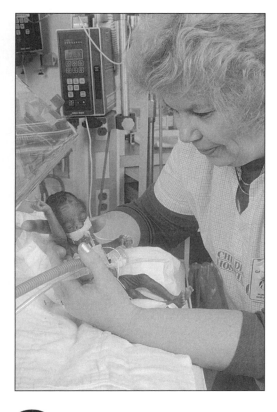

A nurse manipulates a ventilator tube for a tiny premature baby.

PUBLIC-POLICY AND LEGAL ISSUES Medical decisions regarding infants vary in the seriousness of their consequences for families, health providers, society, and the infants themselves. They range from decisions about home birth and circumcision—debatable but generally agreed to be matters of private choice—to vaccination, genetic screening, female genital mutilation, and artificial life support for a critically ill newborn. In the United States, parents' legal right to choose even the most invasive treatment—or to refuse lifesaving measures—was nearly unquestioned until the 1990s. This right has become the subject of lawsuits, extensive scholarly comment, and public concern.

Health-Care Providers' Interests

midwife a woman trained to help other women in childbirth

Historically, treatment decisions rested with the **midwife** or doctor caring for the newborn and its mother. Although parents seemingly "owned" their children, they routinely yielded control to the health-care provider. During the twentieth century, the parent and the provider began to make decisions about medical treatment together. The parents' role has noticeably increased as a result of a greater number of treatment options and increased parental knowledge and awareness.

The physicians' role. Doctors readily acknowledge the frequent conflicts between their two commitments to save lives and to relieve suffering. In reality, these factors are rarely the only ones that affect the doctor treating a critically ill infant. Health-care providers may have differing philosophies about treating infants afflicted with certain disabilities. They may also be influenced by their research agendas, lack sufficient knowledge to assess accurately the infant's disability and prognosis, or be influenced by the risk of legal liability. In addition, doctors tend to focus on the diagnosis rather than on the prognosis and long-term care of their infant patients. As a result of all these factors, doctors may not be the most effective partners in the decision-making process. An **obstetrician** may act in an overly protective way toward the parents, seeking to shield them from the tragedy of dealing with an impaired infant. Alternatively, a pediatrician may be overly optimistic in judging and discussing with the parents the infant's potential for meaningful life.

obstetrician a doctor who specializes in childbirth

Nurses and midwives. Frequently, nurses serve as the main information pipeline between doctors and parents. But nurses have their own biases. Because they are the health-care providers who care for patients most intimately, they may become so attached to a severely disabled infant that they give parents unrealistic hope.

Sometimes, doctors and hospitals overtreat an extremely sick infant to avoid legal liability. In these cases, health-care providers insist on treating an infant without taking into account the best interests of the child and the family. They act simply to protect themselves.

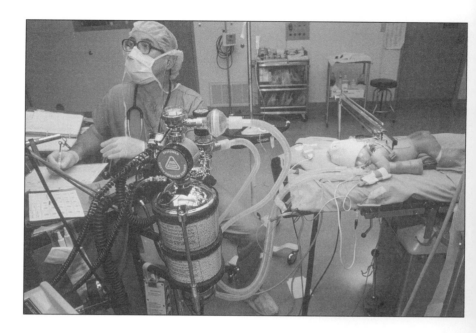

A newborn baby is about to be operated on. The cost of treatment for severely ill infants may deter insurance companies from covering the costs of care for these children.

Society's Interests

A society such as that of the United States has many, sometimes contradictory, interests in the health care of infants. These interests include preservation of the life and health of the next generation; the guarantee of rights to individuals; the support of families; the conservation and wise use of economic resources; the maintenance of a fair legal system; and the compromise of clashing values of groups within society.

Cost of care. Concern for the cost of neonatal intensive care—the most expensive element in the care of infants—preceded the intense focus on health costs in general. Many families cannot cover the cost.

Whether or not cost should affect treatment decisions, there is evidence that it does. Although providers may not abandon a patient without running risk of liability, a study comparing medical need to the services sick newborns receive suggests that health-care providers do not distribute services solely according to need. Instead, they are influenced by the newborn's insurance coverage—private, governmental, or none. Governmental insurance is less attractive to providers than private insurance because the government does not pay back the full cost of care. Thus, at times it appears that while society insists on extending the life of premature and seriously ill infants, it simultaneously refuses to absorb the cost of their care.

Role of the courts. A second important issue is whether society has erred by assigning this category of treatment decisions increasingly to the courts. The judicial system may be too slow-moving and costly and may further traumatize family members and invade their privacy. The publicity surrounding infant-care cases may prevent other parents from exercising their right to refuse treatment. In addition, the practice of medicine itself suffers. Explicit direction from some courts to prolong life whenever possible and the implicit threat of lawsuits reinforce medicine's tendency to overtreat.

Conclusion

Long-standing respect for parent and health-care provider discretion in making infant-treatment decisions has been partially replaced by a greater emphasis on the rights of the infant. As a result, the roles of the infant, parents, health-care providers, and society in making these difficult decisions are undergoing reexamination. Although infant-care review committees occasionally serve as forums for such decision making, a number of these cases continue to be referred to the courts, where little, if any, consensus has emerged.

CHILDREN

This entry consists of two articles explaining various aspects of this topic:

Health-Care and Research Issues
Mental-Health Issues

In 1959, the United Nations General Assembly adopted a ten-principle Declaration of the Rights of the Child. Since then, philosophical interest in the rights of children has grown considerably. This growth owes much to the social upheavals of the 1960s and 1970s, especially the civil rights and women's movements, both of which used the **rhetoric** of rights. When the plights of children began to be highlighted, it was natural that their adult supporters also used the rhetoric of rights.

The key features of the rhetoric of basic rights are (1) that rights are **entitlements**, and (2) that they impose duties on others. To claim something as a fundamental right is to make the strongest kind of demand. To do so is to claim that something is an entitlement, not a privilege—something it would be wrong and unjust to withhold. And if one person is the bearer of rights, some or all others are the bearers of duties.

rhetoric the art of speaking or writing, especially the persuasive use of language to influence the thoughts and actions of listeners

entitlement a legal right or claim

HEALTH-CARE AND RESEARCH ISSUES The most important aids to help children grow and develop well are peace, good parents, healthful food, decent housing, and cleanliness. It is generally agreed that children should also have a way to get basic health-care and social services.

pediatrics a branch of medicine dealing with the development, care, and diseases of children

In the United States, health care as a whole absorbs nearly a trillion dollars annually, but most of these dollars escape child support services. While the health-care system is so vast that cost containment is necessary, children's protective services are still under-funded and, therefore, fail to provide the necessary support to all forms of child care, which should follow the medical model of primary, secondary, and tertiary care.

Stages of Life: Adolescents

Child protective services are often criticized for their policy of trying to keep intact families that have collapsed under the weight of drug abuse. Such a policy means that children are often left to languish in foster care when they could be freed for adoption.

custody guardianship

At the beginning of the twentieth century, the first groups to speak up for neglected and abused young people were home-health nurses and activists in the women's rights movement. The newly formed medical specialty of **pediatrics** also supported these ideas. As the century progressed, lawyers and social scientists joined this movement. They argued against the old idea that children were the property of their parents or guardians. Indeed, under outdated laws, the state had no right to step in even if the children were abused or neglected.

Children soon gained rights to some medical services and to be protected from abuse, poverty, and neglect. Adolescents won the right to consent to some kinds of medical treatments and services without getting permission from their parents. Scientists also helped change certain programs by studying the growth and development of children. They also looked at the illnesses of young people as well as their needs, experiences, and viewpoints.

Who Has Authority to Decide for Children?

Ideally, important health-care decisions that concern a child should be made by a group of people, including parents, doctors, nurses, and the child (if the child is mature enough and wants to participate). Together they should try to find the best solution for the child and the family. Legally, however, parents or guardians have the responsibility and authority to make medical decisions for children.

Parents and guardians as sole decision makers. This thinking comes from the same ideas that say that those caring for children can choose their religion and schooling. First, parents and guardians know their children best and are interested in them. They are, therefore, most likely to do the best job for them. Second, the family is usually responsible for the choices made for the child. Families may be more comfortable with certain choices and not others. Third, children learn values and ideals within their families. Different values and ideals may lead to different health-care choices. These family standards should ordinarily be honored because it is within the family structure more than anywhere else that people in society learn values. Fourth, family members need to be close to one another with little or no interference from the state. So, unless the children are at risk, the families should be given fairly broad decision-making power in selecting health-care options. Parents or guardians will continue to have this authority as long as they help to maintain the health of and decrease the risk of harm to their children.

Parents who abuse, neglect, or take advantage of their children may lose **custody** of them temporarily or permanently. Sexual, physical, or emotional abuse inflicted on their children constitutes legal grounds for the parents to lose authority over them. Parents who make unwise or neglectful decisions may also end up in the same position. Parents, for example, may decide to refuse standard medical antibiotics treatment for a child's infection. If the parents chose instead to use an unusual treatment, such as herbal teas, they could temporarily lose the right to make decisions for that child. When a parent's acts put a child in danger, then doctors, nurses, hospital staff, and social workers have a moral and legal duty to get a court order to see that the child is cared for properly.

Article 25 of the *Universal Declaration of Human Rights*, adopted by the General Assembly of the United Nations in 1949, reads as follows:

(1) Everyone has the right to a standard of living adequate for the health and well-being of himself and his family, including food, clothing, housing, and medical care and necessary social services, and the right to security in the event of unemployment, old age, or other lack of livelihood in circumstances beyond his control.

(2) Motherhood and childhood are entitled to special care and assistance. All children, whether born in or out of wedlock, shall enjoy the same protection.

Informing the children. Another issue is when to tell children about their health-care choices. This seems to be more important in the cases of serious illness where different treatments may get different results. Some children want to understand the decisions about their health care, while others do not. Older children will often have an opinion about how they want their care managed.

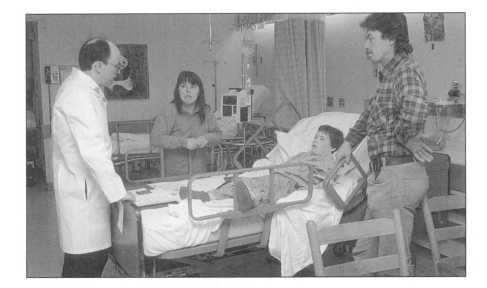

A pediatric surgeon explains treatment to a boy and his parents before surgery.

This trend to keep children more informed and to allow them to participate in the decision-making process comes from several areas. Research has shown that children of different ages and stages of development understand situations to various degrees. Social-science research has found that many children understand much about their disease. They also seem to understand the possibility of their own death. They sense when people are not truthful, and this can hurt them by making them feel alone and powerless to form decisions that will affect them. When children are able to understand and are properly prepared, telling the truth usually increases cooperation and helps to build trust in the health-care team.

In general, children are increasingly competent to take part in health-care decisions as they become better able to understand their long-term options for living. While young children cannot do this, some adolescents are as able as most adults to take part in their own medical decisions.

Children as Research Subjects

In order to grow and to develop to their full potential, children may need social services that allow them to develop their talents and to become independent and happy people. Good health care is the result of research and study. The problem is how this research should be done to help children.

The ethically based research policy with children would contain the same basic values that shape treatment decisions—increasing well-being and opportunities. Small children, like adults severely impaired with mental illness or retardation, lack the ability to give informed consent. This could make them the victims of some research projects. Four important options for this policy have been presented. Each takes a different point of view.

1. The "surrogate" or "libertarian" solution allows the same kind of research with children as with adults, if the parents give consent. This solution may not offer enough protection to children because it permits parents to enroll them in potentially harmful research. One assumes that the moral and legal responsibility of the parents will ensure the safety of the children. Parents have no authority to enter their children into potentially harmful research projects. This kind of action could violate their protective role.

2. The "no consent–no research" or "Nuremberg" solution does not allow children to participate in research because they are not considered capable of consent to participation in research studies. This view, originally stated in the Nuremberg Code (1947), appears too restrictive. It forbids enrolling a child in a study even if the project could directly benefit the child.

see also

Ethics and Law: Information Disclosure, Truth-Telling, and Informed Consent

A nurse reads a book to a young patient undergoing cancer treatment.

3. The "no consent–only therapy" or "Helsinki" solution holds that individuals who are unable to give informed consent may be allowed to take part only in therapeutic studies. This view, found in the World Medical Association's Declaration of Helsinki (1991), is controversial because the definition of a "therapeutic" study is too loose. There are other problems as well. Therapeutic studies often include such nontherapeutic features as lab tests, hospitalizations, or doctor visits. Patients in nontherapeutic research can receive other benefits, such as access to special care. Some nontherapeutic studies that put children at little or no risk—such as those that identify at what age children are mature enough to name animals—may provide important information.

4. The "risk–benefit" or "U.S. federal regulation" solution allows research with children as long as the risk of harm, discomfort, or inconvenience is low and the chance of benefit is high. There must be a balance between protecting the child and the need for research. As the risk becomes greater, the protection and information given with consent become more strict and detailed. This solution is favored by many countries, such as the United States, Canada, the United Kingdom, and Norway.

Human Growth Hormone

The American Academy of Pediatrics Committee on Bioethics in 1997 issued guidelines on the use of human growth hormone (GH) in children. Children taking GH must receive needle injections two to three times per week. Parents want their children to grow, but how much and for how long? School-based screening of children for short stature raises questions about the identification of candidates for GH treatment, and about an unjustified social bias against people who are on the short side. Risks include the psychological trauma of years of regular injection and increased negative self-image due to shortness. The American Academy of Pediatrics urges pediatricians to exercise caution in administering GH hormone.

It is difficult to say when research can be allowed even if its goals are to make a child's life better. On the one hand, if the research is not conducted with children as subjects, then children may not get the benefits of this work. They could miss out on good information about procedures that could promote health and prevent, treat, or diagnose disease. On the other hand, allowing children to be research subjects amounts to using individuals who can be easily harmed and who cannot give informed consent.

In *Amazing Grace: The Lives of Children and the Conscience of a Nation* (1995), Jonathan Kozol describes the lives of children in one of the poorest sections of the South Bronx. Kozol is pessimistic about people of wealth ever helping to build a "new society," and he thinks that the "children of disappointment will keep dying."

With the Adoptions and Assistance Act of 1980, the federal government stated its preference for family preservation. Where earlier policies encouraged swift removal of children at risk to foster care and to adoptive parents, the 1980 federal law requires child welfare workers to make "reasonable efforts" to keep children with their birth parent(s) whenever it is "safe" to do so. Instead of taking the children away, social workers must devise ways to prevent abuse and neglect within the parents' homes.

Resource Allocation

Many children throughout the world do not receive basic health-care or social services. Sometimes, countries that can afford to do so provide inadequate funds for this purpose. The main health problems of children in the United States arise from poor funding of programs that provide basic care for children's allergies, asthma, dental problems, hearing loss, bad eyesight, and many ongoing disorders. These basic services promote children's well-being, increase their chances to succeed, and correct some of the unfair natural and social issues they must face. Children who are ill cannot compete with those who are not and are at a great disadvantage. As it gets easier to treat these conditions, it becomes more unfair to leave the children sick and disabled.

Failing to provide children with basic health-care and social services, when a society has the means, is universally recognized as unjust. This general agreement serves as a powerful indictment. It proves that, as a matter of justice, we should redistribute goods, services, and benefits more fairly to children in order to provide them with basic health-care and social services.

A just society owes all its children basic health-care and social services to help change unfair situations and allow them to develop as free people who can reach their full potential as individuals. Children living in low-income homes in the United States are two to three times as likely as children in high-income homes to have common childhood problems. Poor children are also three to four times as likely as rich children to become ill and to develop multiple illnesses when they get sick. This gap between the rich and the poor is increasing. Rising health-care costs are making it more difficult to distribute services fairly. Since children need others to defend their rights, adults must continue to set aside their own interests and give priority to children's well-being, needs, and opportunities.

MENTAL-HEALTH ISSUES Until the 1970s, many professionals were not sure that children could experience "genuine" depression. It has become clear that children do experience depression, but show it much differently than do adults. Usually, children are much less likely to talk about their feelings. They often hide their symptoms in order to avoid embarrassment, or to keep from upsetting their parents. They also sometimes try to "protect" adults who do not seem to be able to tolerate a child's sadness. Some children may not want to attend therapy if it interferes with their normal daily activities.

One of the important breakthroughs in understanding the mental health of children has been the recognition that what children tell us depends on what we are willing to hear. That is, children will let adults

A homeless family on a street corner asking for help. Homelessness is a real threat to the mental health of children.

know their feelings in a way that makes sense only if the adults have the skill and knowledge to understand what they are being told. For adults to be able to respond to the mental-health issues of children they need to understand some basic ideas of child development.

Child Development and Mental-Health Issues

The stress of city life associated with family problems appears to be strongest for young children and most negative for their mental health. Still, it can stimulate some adolescents who have had a positive childhood. Research may help us to better understand challenges and problems that aid growth and development. The mental health of children can suffer both from directly negative stress or from the lack of normal opportunities. Homelessness is one example of a risk factor with terrible consequences for a child's mental health.

Children make up stories about their lives in so-called social maps of the world. The child's social maps are both a product and a cause of behavior and development. Situations involving trauma, such as the loss of a parent or home, are especially important in forming these memories and stories. The ability of children to have an idea of where their home is and who their families are creates the basis of these social maps.

The Importance of "Home"

The idea of "home" implies a situation that doesn't change. "Home" means you will always have a place to go, no matter what happens. Home is a place where you are connected forever, a place that will last, a place that tells who you really are. As a young homeless child wrote, "A home is where you can grow flowers if you want."

Some experiences that seem similar at first glance may have quite different effects on a child. Being an "immigrant" and a "refugee" are not the same. Neither are having "moved" and being "displaced." With millions of children worldwide suffering homelessness, this very important issue needs to be looked at more closely. Homelessness is a traumatic situation, and it reminds us of the role that trauma plays in understanding the mental health of children.

Childhood Experience of Trauma

Psychological trauma is the shock and confusion caused by exposure to horrible events. It is an important area of study for those who want to understand mental health in childhood. As with depression, it was once thought that children could not experience real trauma. Since 1980, however, research has found that trauma and **post-traumatic stress disorder** play major roles in the mental health of children.

Children experience trauma in many settings: violence on television, community violence, violence at home, war, homelessness. All point to the need to better understand the impact of trauma on childhood. Research in this area will also help the study of the larger issue of mental-health problems facing children.

The manifestations of trauma. Children may suffer from post-traumatic stress disorder due to their experiences at home, in school, or in the community. Its symptoms in children include disturbed sleep, daydreaming, acts of trauma during play, anxiety, decreased hope in the future, and even chemical changes in the brain. Trauma can produce severe psychological problems that interfere with learning. It can also affect social behavior in the family or school.

The children least able to cope with trauma outside the home are those who experience psychological, physical, or sexual abuse at home. Traumatized children can have memory problems and may misunderstand information. This makes it more important for adults to be able to help children tell their stories. In looking for information from children, we must be sensitive to their need to make a story about themselves and their family that is in keeping with their emotional needs. We must also realize that such a story must be consistent with their age, education, and intelligence.

A population at risk. Hundreds of thousands of children face the mental-health challenge of living with constant community violence. Since the early 1970s, the rate of serious assault in many cities in the United States has increased dramatically, in spite of some decline in the late 1990s. Interviews with families living in public-housing projects in several major metropolitan areas show that almost all the children have had firsthand experiences with shooting by the age of five. Some 30 percent of children living in high-crime neighborhoods have witnessed a murder by the age of fifteen, and more than 70 percent have witnessed a serious assault.

post-traumatic stress disorder
a psychological reaction that occurs after a highly stressful event, usually characterized by anxiety and nightmares

▲ A bruised girl huddles against a wall and conceals her face behind a teddy bear. Children who have had traumatic experiences need to be cared for by trained professionals who know how to help them tell their stories.

Conclusion

The job of dealing with the consequences of these problems goes to the people who care for these children—their parents and guardians, other relatives, teachers, and counselors. But these adults face great challenges of their own. They themselves are exposed to extreme violence in inner cities. Human-services professionals and educators often flee these neighborhoods, making the situation even worse for the children who are left behind. Finding the emotional strength and financial support to fight the battle against these problems is one of the major ethical issues faced by policymakers, professionals, and private citizens everywhere.

Not only do many children experience violence firsthand, they can also be violent themselves. In the period from January 1995 to March 1998 in ten different incidents, boys ranging in age from 11 to 17 killed a total of 20 teachers and fellow students in their own schools.

Related Literature

Robert Penn Warren's story "Blackberry Winter" (1946) describes a day during which a boy discovers the many hard realities of adult life. The story takes place on a very cold June day (hence the title). Many bad things happen to the boy and to others he meets. To the youngster, the hope and promise of spring seem suddenly crushed. At the end of the day, the boy comes to the realization that he can never again see the world as the good, secure place he had thought it was.

ADOLESCENTS

Adolescents are defined as individuals between thirteen and eighteen years old. At the close of the century, adolescents have much greater rights to make health-care decisions than they did as recently as the 1970s.

General Consent to Medical Procedures

Under British common law—a system that the United States kept when it declared its independence—a minor was considered property owned by his or her parents. A "minor" meant someone under twenty-one until 1971, when the Twenty-sixth Amendment to the Constitution was adopted, giving voting rights to eighteen-year-olds. Child abuse was not illegal anywhere in the United States until 1903 because people believed in parents' rights to discipline their children. A doctor who treated a child without parental consent (except in an emergency) could be liable to the child's father for interference with control of his child.

venereal disease any of several contagious diseases, such as syphilis and gonorrhea, contracted through sexual intercourse

Many states give decision-making authority (without parental involvement) to adolescents who, while not emancipated, do have considerable maturity and capacity. This is especially so when the adolescent is seeking treatment for sexually transmitted diseases, substance abuse, and pregnancy. The American Academy of Pediatrics emphasizes the maturity of adolescents 14 years of age or older.

informed consent consent to medical treatment by a patient—or to participation in a medical experiment by a subject—after achieving an understanding of what is involved

Ethics and Law: Information Disclosure, Truth-Telling, and Informed Consent

Changes in the legal environment. Beginning in the early 1960s, doctors urged state lawmakers to recognize that great numbers of adolescents were getting **venereal disease**. Because adolescents knew that doctors were unwilling to treat them without parental consent, they did not seek medical care. They feared, with good reason, that their parents would be informed of their disease even if they received no treatment, so they refused to go to doctors. By the end of the 1960s, all states had passed laws permitting treatment of minors for venereal disease without notifying their parents. In the early 1970s, similar laws were passed allowing treatment of adolescents for alcohol and drug abuse without parental notification. Also, beginning in the 1960s, many states passed laws stating that minors of a given age (from fourteen to sixteen) could consent to medical or surgical care.

The "mature minor" rule. Even before these state laws were passed, courts had stopped treating minors like possessions. Not since the 1950s has any court in the United States allowed parents to recover damages from a doctor for treating their adolescent (fifteen or older) without their consent, but at their child's request. Courts in these cases consider an adolescent able to give consent when he or she is mature enough to give the same sort of **informed consent** required from an adult patient. This is known as the "mature minor" rule. The decision to treat an adolescent without parental consent depends not only on the child's age and maturity level, but also on the seriousness of the illness and the risks of treatment. Although most doctors would treat adolescents without parental consent for sore throats or earaches, it is not likely that a doctor would treat adolescents with cancer who refused to involve their parents.

Treatment without parental consent. In reality, doctors' decisions on whether to treat adolescents rarely involve serious illnesses because hospitals, for financial reasons, refuse to admit minors without parental consent except in emergencies. Since those adolescents who are insured are usually covered by their parents' health-insurance policies, the policyholders' consent (that is, the parents' consent) is required for nonemergency situations. Thus issues about treatment of adolescents without parental consent almost always involve outpatient care given in clinics where the doctors will accept whatever payment young people can make on their own.

Even under common law, children of any age can be treated without parental consent in an emergency—that is, anything requiring immediate care, even if not life-threatening. A two-year-old with a broken arm who is brought to the emergency room by a twelve-year-old baby-sitter would be treated even if the parents could not be found.

"Emancipated Minors." A child considered to be an **emancipated** minor has always been able to get treatment without parental consent. The modern definition of "emancipated minor" includes minors who are married, who are in the military, or who are not living with, and are not financially dependent on, their parents. In some states, minors who live at home but who support themselves are considered emancipated. In almost all states, unmarried mothers (as young as eleven or twelve) are considered emancipated even if they are living with their parents. In some states, pregnant minors of any age are emancipated.

Adolescents who are emancipated under their states' laws may consent to or refuse medical care, buy and sell property, and sue or be sued. Parents of emancipated adolescents are not financially responsible for them, nor do they have to give them any further parental care.

Refusal of treatment. Another issue arises when minors refuse medical treatment that their parents want them to have. When an adolescent's life is not at stake, most courts will permit the young person to decide the matter.

Only one case has reached the appeals courts in which an adolescent, a Jehovah's Witness, refused life-saving treatment. She had leukemia and needed blood transfusions, which are forbidden by her religious beliefs. The trial judge decided that the patient understood that she would die without treatment. He believed that her refusal was based on her religious convictions and not on coercion by her parents or fear of their reaction if she consented to treatment. So, the judge allowed her to refuse treatment. The decision was upheld by the appeals court.

The law does not clearly define the limits of an adolescent's right to refuse life-saving treatment. At least one court has held that an adolescent and her parents could not refuse life-saving treatment for her bone cancer on religious grounds. (The court ordered the treatment provided.) The situation of an adolescent wishing to refuse life-saving treatment is rarely brought before courts. Thus it is not known how well cancer specialists, for example, deal with certain issues that may arise for the adolescent patients involved, such

emancipate to free from parental care and responsibility and give full legal rights

The traditional idea of informed consent clearly applies to patients who have reached the legal age of majority (18 years). Emancipated minors can legally consent without parental involvement. These are adolescents who are self-supporting, married, pregnant, or parents. Not all states have emancipated minor laws.

Scott Rose had Nezelof syndrome, an immunodeficiency disease that required him to live in a plastic "bubble." Scott lived in the bubble until the age of 14, when he decided against it. His doctors were against this decision, but his parents approved. With his lungs deteriorating, Scott disconnected himself from the ventilator that was keeping him alive. Scott died hours later.

as their feelings about their appearance. For instance, how much weight is given to an adolescent's belief that he would rather die than have his leg removed in the course of treatment for bone cancer? Are adolescents' views about such issues respected? Or are they treated over their objections but with parental consent, with no outsiders (such as judges) knowing about it?

Confidentiality. If an adolescent is accepted for treatment without the parents' knowledge, the question becomes what information, if any, the parents are entitled to have. Once the adolescent is accepted for treatment—because the doctor feels certain that the patient can give informed consent—the implication is that confidentiality has been promised. A doctor who is not comfortable with this should not agree to provide treatment without parental involvement.

Most state laws that allow treatment of minors for venereal disease, drug abuse, or alcohol problems without parental consent forbid telling parents without the patient's consent. These laws say that parents cannot even know that the medical condition exists or that treatment has been provided. Some of these laws also forbid sending medical bills to parents, further protecting the child's privacy.

The AIDS crisis presents important questions about adolescent patients' rights to confidentiality. An HIV-infected adolescent may have a life-threatening disease, and the medicine used to treat it (such as AZT) is very expensive, usually well beyond the reach of a teenager without parental support and insurance coverage. But there is much evidence that, as with venereal disease in the 1960s, adolescents will not go to a clinic for testing or for care unless they are convinced that their parents will not be contacted. Long-term studies of patients in New York City indicate that teenagers whose parents do not know they are HIV-positive do as well as those whose parents are involved. In order to get adolescents tested and treated for HIV, some hospital clinics agree to provide free medication if the adolescents do not wish to involve their parents.

In most cases involving a serious medical situation, doctors and nurses urge adolescents to tell their parents and agree to help them do so. Most adolescents confronted with a crisis accept to involve a family member. But for some adolescents, telling their parents is not an option, and most doctors and nurses know this. They also know that adolescents in troubled, unsupportive families will not seek care unless they know their confidences will be protected.

Birth control. In 1965, the U.S. Supreme Court held that state laws forbidding the prescription of birth control to a married woman violated her (and her husband's) constitutional right of privacy. In 1972, the Court applied the same rule to unmarried adults. In 1977, the Court held that minors have the same constitutional right of privacy and that state laws making it a criminal offense to give birth control to minors were unconstitutional.

Title X of the 1970 U.S. Public Health Service Act requires that family-planning services be accessible at federally funded clinics

see also

Ethics and Law: Confidentiality

see also

General Topics: AIDS

see also

Fertility and Reproduction: Fertility Control

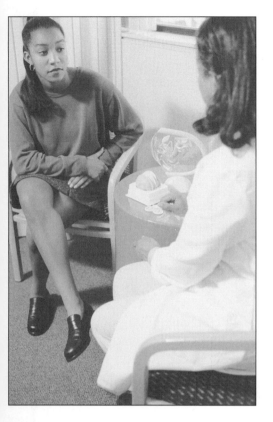

 Discussing birth control.

Fertility and Reproduction: Abortion

General Topics: Abuse

without regard to age, religion, race, or other trait. In 1978, Title X was amended specifically to include adolescents. But in 1981, it was again revised to read, "To the extent practical, entities which receive grants or contracts under Title X shall encourage family participation in projects assisted under this subsection." Federal regulations were then enacted that required any entity receiving federal money for family-planning services to notify parents within ten days that their minor daughter had received birth-control services. But these regulations were immediately declared unconstitutional on the grounds that they violated Congress's intent. So a minor has the right to get birth-control services from a federally funded source. This law does not apply to a doctor in private practice, who then—when asked to give birth control to an adolescent—is legally able to refuse on the basis of ethical objections.

Abortion. Beginning in 1979 with *Bellotti* v. *Baird*, the Supreme Court has held that a state that wants to pass a law requiring a parent's consent to a minor's abortion must allow for exceptions. The minor may bypass parental consent by getting permission for the abortion from a judge. The judge's role is to determine whether the adolescent is mature enough to make an informed decision. If she is sufficiently mature, she has the right to a court order permitting an abortion.

When discussing a minor's right to abortion, it should be kept in mind that an adolescent mother, no matter how immature, has always been considered emancipated for the purpose of giving her baby up for adoption. An adolescent's parents have never had the right to force their daughter to relinquish the baby for adoption. It should also be noted that no matter how young the mother, she has the same legal rights and responsibilities regarding her child as a legal adult has. Her parents have no duty to support or care for her baby—except in the state of Wisconsin, where a law requires the parents to do so until the minor mother is eighteen.

The supposed reason for state laws requiring a parent's consent or a judge's approval for abortion is the belief that young women are "too immature" to make this decision for themselves. If a girl needs parental consent to choose abortion, does she have the right to refuse an abortion her parents want her to have? In the few cases where parents have sought a court order to force an adolescent daughter to have an abortion, the courts have held that the parents could not do so.

Sexual abuse. In 1984, the California attorney general ruled that all sexual activity involving children under fourteen had to be reported as child abuse, regardless of whether the sexual activity was voluntary. Anyone under fourteen who was "treated for venereal disease, birth control, for pregnancy or for abortion" was to be reported as "abused." Planned Parenthood and several doctors sued to challenge the ruling, on the grounds that it invaded the minors' privacy rights. The California Court of Appeals declared the

consensual based on mutual agreement

psychotherapy the treatment of mental or emotional disorders by psychological means

Mental Health: Mental Health and Mental Illness

psychiatrist a doctor who treats patients who suffer from mental illness

ruling unconstitutional. It held that child-abuse laws do not require reporting where there is no knowledge or suspicion of abuse solely because the child is under fourteen, if the minor states that the sexual activity was voluntary, **consensual**, and with another minor of "similar age."

If an adolescent is reported as abused, child-welfare agencies begin investigations with home visits. There is then no chance that a minor's pregnancy or venereal disease (in spite of specific confidentiality laws) will be kept from parents. If the possibility exists that a doctor must report minors as abused simply because of age, minors will be less likely to seek medical care.

Mental-Health Issues

When adolescents seek mental-health care on their own, it is usually in community mental-health facilities, drug treatment centers, counseling centers, or other public facilities. The issue of an adolescent's right to mental-health care is unlikely to involve private **psychotherapy**, since the average young person cannot afford it. Community mental-health centers are generally covered by the same rules that apply to other medical treatment, since those institutions usually receive government money and are careful to comply with all regulations that apply.

Many behaviors that seem normal to an adolescent may seem abnormal to a parent, who may seek psychiatric treatment for the child. Parents may seek such treatment even though the minors are functional, are not engaging in criminal or delinquent behavior, and are not dangerous to themselves or others.

Refusal of psychiatric treatment. If minors have the right to consent to treatment, they generally have the right to refuse it. In any case, psychotherapy is unlikely to work on an unwilling patient. The patient who is forced to participate in psychotherapy can simply refuse to discuss anything.

Confidentiality and psychiatric treatment. Does a **psychiatrist** owe an obligation of confidentiality to a minor patient? There are only a few state court decisions that discuss minors' rights to keep statements they have made to doctors unavailable to parents. A parent and an adolescent may be opposed to each other because of the nature of their relationship. Psychiatrists must often assume that anything they tell a parent will be used against the young person.

Some authorities believe that a psychiatrist is free to—and should—discuss anything he or she chooses with the parents of minor patients, and also with the minors' teachers. This is a violation of any child's right of privacy and may also be poor therapy. Telling a teacher that a student is in psychiatric treatment may cause negative feelings against the student. The teacher might add the information to the student's permanent school record. Minors should have the same right of confidentiality that adults have, except in situations where the patients may be dangerous to themselves or others. So, as with any adult patient, information should be given to parents or others only with the patient's permission.

The psychiatrist's "duty to warn." It is a matter of professional judgment whether an emergency exists that justifies revealing confidences to a parent or someone else.

A parent's right to know is less restricted than anyone else's. It follows that a psychiatrist's "duty to warn" the parent that an adolescent may endanger himself or herself or others can be based on evidence that is much weaker than that required for an adult patient. The psychiatrist's decision to report such information to parents should be based on a lesser probability of harm than that required for the doctor to inform the police, the threatened victim, or anyone else.

Commitment to a mental institution. There are two distinct standards for commitment of adult patients to mental institutions. Adults are committed involuntarily (against their will) when they are "dangerous to themselves or others" or are considered "gravely mentally disabled." (The legal definition of the latter is that as the result of mental illness, patients are unable to provide food, clothing, shelter, and medical care for themselves.) Adults are committed voluntarily when they decide, with their doctors, that in-patient treatment would be beneficial.

Minors of any age are in an entirely separate category. Many state laws allow "voluntary" commitment of children by their parents. Minors who are committed "voluntarily" by their parents have fewer legal protections than adult patients have. Adult voluntary patients in a mental hospital can leave at will unless, after arrival at the hospital, they are classified as "dangerous" or "disabled." (At this point, a judge must hold a hearing and the patient must be committed or allowed to leave.) Involuntary patients have the right to a judicial hearing when they enter the hospital and the right to leave when they are no longer dangerous to themselves or others. Most states will not let minors leave a mental hospital without their parents' approval. Thus hospitalized minors have far fewer due process rights than do adult patients.

Legal Issues in the Treatment of Mental Disorders

Case law indicates that there are many situations in which abusive parents have tried to commit their children to mental hospitals for reasons unrelated to the children's condition. In the 1960s, some male adolescents were confined to hospitals for months or years because they refused to cut their hair. In many cases, parents have sent their adolescents to mental institutions without any serious attempt by doctors to determine if the young people are really mentally ill.

If an adolescent has conflicts with a parent, society often concludes that the young person, not the parent, is the one with the

Mental Health: Abuses of Psychiatry

see also
Mental Health: Commitment to Mental Institutions

Ethicists point out that often, adolescents who have experienced years of repeated aggressive medical treatment for severe diseases suffer so much that it is reasonable for them to decide, at some point, against further efforts to extend life. Their maturity should be acknowledged and their wishes granted.

In the Minnesota case of *Lundman* v. *McKown* (1995), an appellate court overturned a ruling for damages against the Christian Science Church in the case of an eleven-year-old boy who died of undiagnosed juvenile diabetes. His parents had relied on prayer alone to heal him.

problem. This is not necessarily true. Furthermore, a parent cannot be assumed to be acting in the best interests of a child when undertaking commitment proceedings.

Children's rights. In the early 1970s, several cases held that children have certain minimal rights of due process before being committed to a mental institution. Children also have the right to be released from a mental hospital or institution for the mentally retarded on constitutional grounds if they have been denied a fair hearing and a lawyer.

In 1979, the U.S. Supreme Court held that a minor's federal constitutional rights were not violated even though the state did not grant the minor the right to a hearing, a lawyer, and certain due process rights before "voluntarily" admission to a mental hospital by a parent. The Court held that to protect minors, the parents' decision to commit had to be reviewed by a "neutral fact finder." But this fact finder could be a staff doctor, as long as the review of the decision was independent. After this decision by the Court, almost no more states passed due process laws for minor mental patients.

In 1991, the press reported that some profit-making mental hospitals admitted any adolescent patient whose parents sought admission. Some paid high-school guidance counselors money to convince parents that their children needed to be hospitalized. After the adolescents were admitted, the hospitals then refused to release them for weeks or months. The possibility of abuse in this situation is great because, once hospitalized, the patients can be kept from all outside contact. Few state legislatures or judges have been willing to deal with this problem.

Commitment to substance treatment institutions. Another growing problem involves the rights of young people whose parents have admitted them to an alcohol or drug treatment facility. The courts in at least two states have held that since these institutions are not "mental hospitals," any rights under state law to a judge's independent review do not apply. The courts will not question a parent's right to admit an adolescent, even if the adolescent is not diagnosed as "addicted." A minor in such a case who is unjustly hospitalized may have no ability or right to get outside help of any sort. By contrast, if the parent wants to have the child declared "unmanageable" for the same behavior and turns to the juvenile court system, the child has the presumption of innocence, the right to counsel, and the right to a full hearing.

Conclusion

While most parents attempt to love and be responsible for their children, some do not. Health-care providers who treat adolescents try to respect their dignity, their self-determination, and their privacy. But adolescents receive care in a society that often does not fully understand their interests.

Related Literature

Joyce Carol Oates's short story "How I Contemplated the World from the Detroit House of Correction and Began My Life Over Again" (1970) is written in the form of notes for a high school paper assignment. The teenage narrator, a wealthy girl at a private school, traces her experiences and feelings from her impulsive shoplifting and running away from home to her being used as a prostitute in a drug-infested Detroit slum. After she is beaten up by streetwise girls who resent her privileged background, the narrator goes back home grateful for her comfortable life but deeply changed by her experiences. The story explores how some teenagers live in a very different world than their parents do.

AGING AND THE AGED

This entry consists of five articles explaining various aspects of this topic:

Theories of Aging and Life Extension
Life Expectancy and Life Span
Societal Aging
Health-Care and Research Issues
Old Age

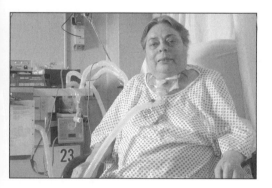

THEORIES OF AGING AND LIFE EXTENSION

Aging is caused by damage to molecules, cells, and tissues. The body's repair processes cannot keep up with the damage, and the organism does not work as well as it did before. The speed of aging is affected by the genetic background of the organism and by its environment. An animal's life span depends on how well its body continues to repair all the damage that occurs over time.

In technological societies, life expectancy has doubled in the last century due to better sanitation, vaccines and antibiotics, and better treatment of disease. But the maximum number of years that a person can live has not changed very much. Even if we eliminate heart disease and cancer, the average life span would be affected only a little, and it would not change the fact that there is a limit to how long anyone can live. If we wanted to increase this limit significantly, we would need to find ways to stop or slow aging.

No one knows how the life span of creatures is determined or how aging evolved. Humans and shrews are both mammals and are closely related to each other compared to other animals like birds, fish, or reptiles. Yet a shrew lives a maximum of one year, while

Everyone wants to die like George Burns, healthy, energetic, and still performing comedy until just months before his death at age 100. Burns is a splendid example of the "compression of morbidity," the ideal of living a very long time but without becoming ill.

senescent the state or process of being old; in cells, a nondividing stage

mitochondria (singular: mitochondrion) long or round-shaped parts of a cell; made up of fats, proteins, and enzymes, they produce energy through cellular respiration

Jeanne Calment (1875–1997) was 122 when she died, the oldest living person on record. She still rode a bike at age 100 and released a rap CD at 121.

people can live to a maximum of about a hundred years, or in rare cases, 120 years. Some people believe differences in animals' life spans depend on how well that animal can repair environmental damage to its body. Aging is not a disease, nor is it a curable condition. It is a set of biological responses to damage. In order to slow or stop aging, we need to understand these responses.

Causes of Aging

Molecular and cellular damage can be caused by the environment or can come from within the animal.

Old cells. Skin cells taken from the average person will divide about fifty or sixty times, enter a **senescent** stage, and then die. If the cells are taken from an older person, they will not divide as many times before they reach senescence. This pattern of dividing a certain number of times and then dying is possibly related to aging.

Mitochondrial damage. Mitochondria produce energy for the cell. This process requires oxygen, but in the process, the mitochondria may produce oxygen radicals. These are a particular kind of oxygen molecule that can cause damage to cells.

The ability of cells to resist, adapt to, or repair damage is a key factor in their survival and in how long the organism can live. Vitamin C, vitamin E, beta-carotene, and other nutrients and chemicals clean up oxygen radicals and help protect the body against them. Also, cells can repair some damage after it happens. With age, these protective processes become less efficient, and aging may happen more quickly.

LIFE EXPECTANCY AND LIFE SPAN In the United States in 1900, the average life expectancy of a newborn baby was 47.7 years—46.6 for males and 49.0 for females. By 1990, the average life expectancy increased to 75.4 years—78.8 for females and 72.0 for males. Why did this increase happen, and can we raise it even more? Did the average person's health improve or worsen during this time? If we could increase life expectancy, how would this affect people's health?

Women live longer than men. In 1900 the average woman lived 2.6 years longer than the average man. By 1990, she lived 6.8 years longer. No one knows why women live longer than men, or whether this gap in life expectancy will continue to widen. Some people think men die younger because they are more likely to have heart disease and cancer, or that men may be genetically programmed to die younger than women.

Whites versus African Americans. There is also a historical difference in life expectancy between whites and African Americans. In the early twentieth century, the life expectancy of African

Americans was ten years less than that of whites. In the late twentieth century, the gap was seven years. Most people believe that this difference is caused by a combination of biological, social, and environmental factors, but no one is certain which factors are most important. Once people have reached the age of seventy or older, African Americans tend to outlive whites.

Extending Life Expectancy

Scientists debate whether or not we can increase the average life expectancy even further. Achieving a longer average life expectancy will affect many social programs, such as Social Security and Medicare, that provide aid to older people. Some scientists say that the average life expectancy cannot get much higher than about eighty-five years. This is because life spans up to or over the age of 110 are very rare and always have been. Also, many people die of aging-related diseases in their older years, and we have not been able to change this. To raise the average life expectancy, we would have to lower death rates at every age, but the sheer number of people and diseases makes this impossible.

Other scientists say that average life expectancy might be increased to 100 years, but this means that everyone would have to live a risk-free life with the perfect diet, exercise, and other lifestyle modifications, which is unrealistic.

Some people believe that improved lifestyles and advances in medicine in the future will keep people healthy until old age, when they will be ill for a shorter time before they die. This is called the *compression-of-morbidity* theory. In this theory illness is compressed into a short time at the end of one's life.

Others believe in the *expansion-of-morbidity* theory. They say that because we have better treatments or cures for the diseases that used to make people die young, more people are living to be old enough to get age-related diseases. This means that more people will experience **Alzheimer's disease**, hearing or vision loss, and they will have more long-term illnesses.

The Issues to Consider

Is it possible to extend the human life span beyond the limits we know, and at the same time allow older people to live free of aging-related diseases? Ongoing research in molecular biology may answer this question. If we can change the genetic programming that causes cells to age and die, or if we can find nutrients or other chemicals that protect the cells, then new questions will come up. How would these advances affect our society? Most people assume that human life is limited. If it were not, what would happen? This question is perhaps much more difficult to answer.

In 1997, the Hemlock Society U.S.A., an organization that supports assisted suicide, issued a statement endorsing the nonvoluntary killing of elderly persons in the last stages of progressive and irreversible dementia. This statement was issued in the context of a Louisiana court case, and in support of David Rodriguez, who was convicted of shooting to death his ninety-year-old father who had advanced Alzheimer's disease.

morbidity the incidence of illness

Alzheimer's disease a degenerative disease of the central nervous system characterized by premature mental deterioration

Demography is the study of population dynamics. Throughout history, the population of all societies has been described as a demographic triangle, with the few very old people at the top, and the many younger people filling out the base. In America's aging society, a demographic rectangle has emerged, with roughly proportionate numbers of relatively old and young people.

SOCIETAL AGING

A society ages when the number of older people increases more quickly than the number of younger people. The societies of the United States and other industrial nations have been aging since at least 1800. At that time, about half the population in developed countries was sixteen or under and very few people lived beyond age sixty. This is similar to the modern societies of many less developed countries, but in developed countries things have changed since 1800. People are living longer and having fewer children, so the proportion of older people is greater. In the future, people may look back at the twentieth century and remark on how fast population aging occurred during that time. People over eighty-five are the fastest-growing age group in the United States. It is twenty-one times as large as it was in 1900, and the number of people who are over sixty-five is eight times greater than it was in 1900.

Ethical Implications

The rapid increase in the number of older people compared to younger ones affects society. It will cause higher health-care costs and more pressure to ration services. This will bring up questions about giving fair health care to both the young and the old. Also, if there are more older people, that means there will be more people with long-term illnesses that are not life-threatening. People often debate immediate, life-and-death medical decisions, but in the future they might debate whether or not someone should get long-term care.

Family life may also change. The younger generation may have to spend more time taking care of their elders. How much of a voice should younger family members have in making decisions about health care for their elders?

Health-care rationing. The aging of society will force us to spend more money on health care, because people over sixty-five require more health care than other age groups. In the United States, people sixty-five and older make up 12 percent of the population, but they use one-third of the country's total health-care spending. Because of this, care for the aged is often a target for budgetary cuts. For instance, if people who are fifty-five and older were not allowed to have treatment for kidney disease, it would save 45 percent of the costs of the government program that pays for this treatment.

In his book *Setting Limits* (1987), David Callahan agrees with others who say that government-funded health care should be rationed. This means that not everyone would get life-extending health care. People who are too old or too sick to be cured might be left out, but would be given better long-term care support. People who believe in rationing say that society has a duty to help everyone live a natural life span, but not to try to keep people alive indefinitely.

"In the future we may have to ration health care."

—Albert A. Gore, Jr., Vice President of the United States, 1990.

Critics of this view say that it violates basic moral values of Christianity, Judaism, and other religions, and that it promotes the idea that old people's lives are not worth much.

Long-term care. The aging of society will also increase the number of disabled people and their need for long-term care. First, the number of older women compared to older men is expected to increase, and older women are more often disabled than older men. Second, the number of people over age eighty-five is increasing most quickly, and this age group is the heaviest user of long-term care. More than 70 percent of people who are eighty-five and older need some kind of help with daily living.

Fewer children will be born to help take care of older people, because more women are working, and fewer couples are having children.

The growing need for long-term care brings up questions about dividing up health-care money fairly. Should it be used for long-term care, or for short hospital stays?

HEALTH-CARE AND RESEARCH ISSUES Older people usually are less healthy than younger people and are more likely to have to make difficult decisions. Should they be put on life-sustaining machines if necessary? In the United States, many people do not die until after they have made a decision to limit their treatment. When this happens, people receive pain relief, food, water, and other basic necessities, but they are not put on life-support machines.

Advance directives. Older people are more likely to have problems, such as Alzheimer's disease, that prevent them from deciding capably. When this happens, their families often have to

An Alzheimer's patient receives a visitor in a long-term health-care facility.

Alzheimer's disease (AD) affects about half of people over eighty-five years old, and many people in their seventies and early eighties as well. It destroys brain cells and thus devastates all normal human functioning. Of the 1.8 million Americans currently in nursing homes, about half have Alzheimer's. It has been called "the disease of the century," and will certainly be the disease of the twenty-first century.

In his book *The Moral Challenge of Alzheimer's Disease* (1995), Stephen G. Post emphasizes that caring for elderly parents with chronic diseases that have no cure is a considerable burden on their adult children. A century ago, it was easier to honor father and mother because they usually did not live into extreme old age. But in a "graying society," adult children often have much heavier duties.

make these difficult decisions for them. People who have studied the decisions families make have found that they often do not choose what the patient would have wanted.

More people have become interested in advance directives. These are documents that say what kind of health care the person will want, in case he or she is unable to speak or think clearly at some future time. In 1991, the Patient Self-Determination Act was passed. This law requires health providers to educate patients so that they can make informed, thoughtful decisions about their care. Providers also have to tell patients that they can make out advance directives for their own care. It is hoped that more older people will use advance directives, but it is still too early to tell whether they will, or whether doctors will follow the directives patients give them.

Research on aging. Research on older people is a relatively recent project. The National Institute on Aging (NIA) was established within the National Institutes of Health in 1974 to promote research on aging. That this organization was necessary is clear, because little research had been done on older people, and they were usually left out of studies of common diseases like cancer, heart disease, and high blood pressure—even though older people have these diseases more often.

Various organizations and authorities have created ethical guidelines for doing research with people who cannot think clearly, such as those with Alzheimer's disease.

OLD AGE Most people think that in the past, the elderly were treated with more respect than they are now. In the late twentieth century, many people are thinking about this. People worry that the elderly are using up most of the government's health-care money; there seems to be more conflict between the generations, and medical technology is keeping people alive for longer periods of time. Many also worry that old people are not being treated well. These ethical conflicts and questions seem to recall the biblical plea: "Do not cast me off in old age, when my strength fails me and my hairs are gray, forsake me not, O God."

By the mid-1970s, increasing length of life, economic security, and medical care were all available to more older people because of the government's welfare policies. Soon after this, economic troubles led some people to question spending so much on the elderly. These critics said that more money was going to senior citizens than to families and children, and that this was unfair. People were also concerned about the environment, fears of nuclear war, economic decline, and social conflict.

The fairness issue in spending between generations seems to worry most Americans. Unlike some other cultures, the modern U.S. culture does not hold the elderly in any kind of special respect.

The abstract expressionist artist Willem de Kooning (1904–1997) painted his way through much of his struggle with Alzheimer's. Various art critics commented that his work, while not what it had been, was nevertheless impressive. As one art critic wrote, nearly to the end of his life de Kooning knew what he loved best—art—and it sustained him.

There is no dominant idea that older people have much to say or teach or that they should be respected. Instead, aging is viewed as a medical problem, ending in death, but without much other meaning or purpose.

In the late twentieth century, people are searching for ideals and roles for older people. A larger fraction of the population is thinking about the moral and spiritual aspects of growing old. What people decide, and how they integrate their values with modern medicine and the American idea that each individual should have unlimited development, will affect the answers to many ethical questions about aging and our aging society.

Related Literature

Eudora Welty's short story "A Worn Path" (1941) tells of Phoenix, a one-hundred-year-old African-American woman who walks miles through woods and across farms along a back country path that takes her to town so she can get medicine for her grandson. Phoenix, whose name suggests new life born from ashes, keeps on going in spite of her frail old body and bad eyesight. She encounters many obstacles in her path, but overcomes them or maneuvers around them. The worn path also serves as a metaphor for her life.

Ernest Hemingway's novella *The Old Man and the Sea* (1952) helped earn him the Nobel Prize for literature. The hero is an old fisherman, Santiago, who has accumulated much wisdom in his many years. Santiago is dismissed by his village as a former champion who has run out of luck and energy. He goes out fishing alone, because no one will go with him. After a three-day struggle, he catches a giant marlin. The eighteen-foot fish is much too big to get into Santiago's boat, so he ties it to the side. On his way back home, he battles sharks which eat most of the marlin, but the head, skeleton, and tail that remain tell the villagers how badly they had misjudged the old man's ability.

Therapies

ALTERNATIVE THERAPIES

culture the customs, beliefs, attitudes, and outlook that a particular ethnic, national, or religious group exhibits or espouses

In December 1997 the editors of the *Journal of the American Medical Association (JAMA)* sent out a call for papers on the topics of "complementary, alternative, unconventional, and integrative medicine." This decision was based on a survey indicating that physician readers of *JAMA* ranked alternative medicine as the seventh most important of seventy-three topic areas.

physiology the branch of biology that is concerned with the study of bodily processes, specifically those of the human body

pharmacology the branch of science concerned with drugs and how they affect bodily and mental processes

Has your family ever relied on herbs or teas to help you get well? Are these remedies always good for you? The answers to both questions will largely depend upon the **culture** of the society or group an individual comes from, because every culture has its own beliefs about the nature of healing and the methods that should be used to promote healing.

Thus, systems of healing—that is, medical systems—each tend to have clear, defined views of the right and wrong ways to treat disease and promote healing. Such a set of views is called an orthodoxy, and medical orthodoxies are as common as religious and political ones. An orthodox therapy follows the practices and beliefs of a culture's dominant medical practitioners; an unorthodox therapy does not—its diagnosis or treatment of illness differs from accepted belief and practice. In Western society, and especially in the United States, it has become customary to call such therapies "alternative therapies."

Although many alternative therapies exist, they all have one thing in common: the powerful scientific and medical establishment disapproves of them and expresses its disapproval in economic, legal, and cultural ways. Medical orthodoxy in Western society long ago linked itself to the scientific traditions of chemistry, **physiology**, and **pharmacology**. In so doing, the Western tradition, when discussing disease, chooses to reject the validity of any nonphysical cause, explanation, or treatment. The tradition considers the worth of all healing theories and treatments to be dependent on scientifically proved results—results that can be described, evaluated, and statistically validated. This reliance on evidence links scientific medicine with the economic and legal institutions of modern Western societies.

Nineteenth-Century Alternative Medicine

Samuel Thomson (1769–1843), a poor New Hampshire farmer whose mother and wife had suffered from the bleedings and mercury-based drugs forced upon them by regular doctors, devised one of the first systems of alternative therapy.

Thomsonian system. He studied the healing properties of herbs (sometimes called botanicals) and soon developed a system of botanical medicine based on the assumption that cold was the sole cause of disease and heat the sole cure. He believed that restoring heat to a patient's body would cure any ailment, and so he rejected bloodletting and drugs in favor of steam baths and herbs, such as cayenne pepper (hot red pepper—some people sprinkle it on pizza).

Yellow arnica flowers, members of the sunflower family, are often used in homeopathic and herbal medicine.

The Thomsonian system reached the height of its popularity in the 1820s and 1830s. Some sources estimate that as many as a million Americans used some or all of its methods. During these years, while orthodox medicine was growing more organized and professional, Thomson's method strengthened the role of parents—especially mothers—in caring for family members. The system also fit in well with the religious climate of that era, which urged individuals to take responsibility for their own moral and spiritual improvement.

Homeopathy. As the public gradually lost its enthusiasm for the Thomsonian system, a second form of alternative medicine, homeopathy, emerged. This system was the creation of the German doctor Samuel Christian Hahnemann (1755–1843), who grew increasingly critical of the unrestrained manner in which doctors of that period prescribed drugs. He made up a term—"allopathic"—that he used to describe what he saw as orthodox medicine's over-reliance upon such "invasive" treatments as bloodletting, surgery, and the administration of strong drugs. In contrast to allopathic medicine, Hahnemann developed a theory that relied more upon the body's natural powers to bring about recovery.

The first principle of homeopathic medicine is "like cured by like." By this Hahnemann meant that doctors should treat **symptoms** by prescribing drugs that produce similar symptoms in a healthy person. The second underlying principle of homeopathic medicine is the doctrine of **infinitesimals**. It was Hahnemann's strong belief that the greatest curative benefit was achieved by giving **diluted** doses of a drug, sometimes as little as one-millionth of a gram. Although the use of infinitesimal doses by homeopathic doctors surely lessened or undid any healing value their drugs might have had, at least these small doses had the virtue of not doing harm to the patient's powers of recovery. People turned to homeopathy as a useful and appealing substitute for orthodox medicine.

Throughout the nineteenth century, 10 to 12 percent of U.S. medical schools and their graduates believed in the value and usefulness of homeopathy. Unlike the nonprofessional practitioners of the Thomsonian system, homeopathic practitioners were educated professionals who often came from the ranks of orthodox doctors.

Hydropathy. In the mid-1840s, another alternative therapy began to attract a following in the United States. It was known as hydropathy, which means "water cure." Based on the theories of Vincent Priessnitz of Austria, hydropathy focused on enhancing the body's own vitality and purity. Priessnitz believed that pure water could be used to flush out impurities and stimulate the body's natural tendencies toward health. Treatments emphasized drinking large amounts of water and applying water to the outside of the body through baths, showers, or wet sheets wrapped around the body.

Most American believers in the water cure favored an approach to health that stressed the healing powers of fresh air, good

symptom something that indicates disease or bodily disorder

infinitesimal an amount so small that it can hardly be measured

dilute weaken the strength, flavor, or brilliance of, by mixing with something else

In the United States, the estimated 425 million visits to unconventional medical practitioners in 1990 exceeded that year's total number of visits to primary-care physicians. In 1990, Americans spent an estimated $14 billion on alternative therapies.

Because well-designed scientific studies establishing the effectiveness of alternative therapies are usually lacking, the National Institutes of Health (NIH) established the Office of Alternative Medicine in 1992. The office funded a number of research centers across the country. As of 1998, some NIH leaders still insist that the Office of Alternative Medicine should be closed, but others advocate expansion.

diet, sleep, exercise, and proper clothing. The language used by those who favored the water cure also had a strongly moral tone.

Dietary regimens. Sylvester Graham (1794–1851) combined conservative religious beliefs with a passionate concern for health reform. Graham—who was an ordained Presbyterian minister and traveling preacher of the Christian faith—believed that human physical, spiritual, and moral well-being required careful adherence to the natural order established by God. He urged his followers to avoid alcohol and overindulgence in sexual arousal in order to maintain moral and physical health. His notion of a healthy diet included a coarse bread, later produced in the form of a cracker that is still called the "graham cracker." Graham's dietary principles, widely circulated throughout the nineteenth century, aimed at keeping the soul's "bodily temple" free from impurities.

Seventh-Day Adventists. Ellen White (1827–1915) occasionally visited a resort in Danville, New York, where she became a convert to Graham's dietary beliefs. White later claimed to have had a series of mystical visions in which God revealed to her that he expected humans to follow certain divinely given laws governing health and diet as faithfully as his moral laws. The religious denomination founded by White, the Seventh-Day Adventists, has adopted vegetarianism and Graham's dietary principles as essential parts of a purification regimen in expectation of the second coming of Christ. This denomination, which is one of the largest religious groups founded in the United States, supports a number of treatment and rehabilitation facilities in which their strong emphasis on healthy dietary practices is combined with the preaching of their religious faith. While this emphasis upon a healthy diet is not, strictly speaking, an alternative therapy, their dietary concerns are closely connected with their belief in the healing power of prayer.

Mind Cure and Christian Science

Another popular alternative therapy became known as the mind-cure, or New Thought, movement. Mind-cure writers in the United States published books and pamphlets that described how thought controls the extent to which people are able to make themselves internally receptive to spiritual energies. From the late 1800s to the late 1900s, large numbers of Americans have shown remarkable enthusiasm for literature emphasizing the "power of positive thinking" written by such men as Phineas P. Quimby, Warren Felt Evans, Norman Vincent Peale, Norman Cousins, and Bernie Siegel. The mind-cure movement gave rise to a new form of religious piety based on the belief that the deeper powers of the human mind control access to a power beyond physical nature that can instantly help people achieve peace of mind, improved health, and an endless flow of energy. The **holistic** health movement, which began in the 1960s and 1970s, relies heavily upon this cluster of ideas.

Lundman v. *McKown*, 1995: In Minnesota, an appellate court overturned a lower court ruling that imposed punitive damages against the Christian Science Church. In this case, an eleven-year-old boy died of undiagnosed juvenile diabetes. His mother and stepfather had relied on prayer instead of going to see a doctor. The appellate court determined that neither the mother, stepfather, nor church had to pay compensatory damages.

holistic concerned with the whole body, not just a part of it, with the whole person, both mind and body; and even with the person within his or her immediate environment

At least thirty-four U.S. medical schools offer courses in alternative medicine. A 1995 study of family physicians indicates that more than half consider alternative medical interventions such as biofeedback, hypnotherapy, and massage therapy to be "legitimate medical practices."

In 1862 Mary Baker Eddy, in physical and emotional distress, arrived at the door of the famous healer Phineas P. Quimby. Quimby's treatments gradually cured her of her ailments; they also gave her a new outlook on life, one based on the principle that thoughts determine whether a person is inwardly open to or closed off from the creative activity of spiritual energy. Soon after Quimby's death, Eddy developed his teachings into the foundational principles of Christian Science. Her principal work, *Science and Health with Key to the Scriptures* (1875), reveals her intention to modify the "science" of mental healing so that it would bear a closer resemblance to her own interpretation of the meaning of the Christian Scriptures. The basic premise of Christian Science is that God has created all that is, and all that God has created is good. Thus, sickness, pain, and evil are not creations of God; therefore they do not truly exist. Christian Science healers, known as practitioners, help individuals regain their health by overcoming their faulty thinking and elevating their mental attitudes above the misleading guidance given by the senses.

Osteopathic and Chiropractic Medicine

Osteopathic and chiropractic medicine provide interesting evidence of what can happen to alternative understandings of healing in an age dominated by scientific medicine.

Osteopathic medicine. Osteopathic medicine developed from the philosophy of Andrew Taylor Still (1828–1917). A former believer in other alternative approaches, Still devised techniques for manipulating vertebrae (the small bones that compose the spine) in ways that he thought would remove obstructions to the free flow of the "life-giving current" that promoted overall health. He understood the healing principles of osteopathy—a term drawn from two Greek words meaning "suffering of the bones"—as being spiritually based, as were the nature and origin of the life-giving current, which was ultimately responsible for human well-being. His followers, however, largely ignored the nonphysical aspects of Still's thought. They focused instead on grounding osteopathic medical education in anatomy and scientific physiology. Thus, osteopaths, who had originally relied solely upon spinal manipulation to restore health, soon added surgery and even drug therapy to their medical practice.

By the 1950s so few differences existed in the training and practice of osteopaths and M.D.s (orthodox physicians) that the competing professional organizations of which they were members agreed to end their rivalry and cooperate in such matters as access to hospitals and residency programs. Osteopathy in fact gave up the alternative outlook of its founder and no longer possessed any unorthodox characteristics; it became part of the medical mainstream.

In the 1960s—perhaps reacting to the "mainstreaming" of osteopathy—many osteopaths brought renewed attention to the

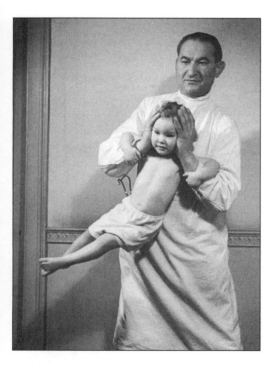

An osteopath treats a young girl by swinging her from side to side.

▶ The crew of a British submarine undergo sun ray treatment, 1940.

The American Academy of Pediatrics is especially concerned about parents who bring their children to alternative therapists whose methods are unproven, especially when proven standard medical interventions are available. The Academy believes that parents who do not bring their children to the pediatrician or to other trained physicians as needed should be held criminally negligent.

▲ A folk healer places hot glass jars on a patient's back (Vietnam).

spiritual aspects of the movement's beginnings. Their focus on osteopathy's historical desire to enhance the body's natural powers of healing made them advocates of holistic medicine a full twenty years before "holistic" became a common term among alternative healers. In 1990, there were over twenty-four thousand osteopathic practitioners, who together treated over twenty million patients a year.

Chiropractic medicine. The case of chiropractic medicine is more complex. Chiropractic began with the work of Daniel David Palmer (1845–1913), who had been a mind-cure healer in Iowa. Palmer, who knew of Still's osteopathic techniques, came to think that dislocations of the spine could block the free flow of the "life force," which he called the Innate. He and his son, B. J. Palmer, explained that Innate is part of the "divine intelligence" that fills the universe, bringing full physical health whenever it flows freely through the human body. Chiropractic medicine represents the Palmers' art and science of adjusting the spine in ways that remove obstructions to the free flow of Innate within the body.

As time went by, chiropractic doctors laid less stress on the movement's origins (much as osteopaths had done) and instead emphasized its scientific approach to the treatment of muscular and skeletal disorders. So they both minimized the unorthodox aspects of their theory and identified an area of medical concern largely ignored by orthodox doctors.

Even though most insurance companies now recognize the medical functions performed by chiropractic medicine, M.D.s remain reluctant to admit its value because it does not have a cohesive system of thought to support its healing therapies. This

A woman meditating and holding a crystal.

metaphysical relating to matters beyond the perception of the senses; supernatural

professional tension is a good example of a continuing theme in the history of alternative medicine: the clash between orthodox medicine's rationalism (its insistence upon a scientific explanation for all treatment methods) and alternative medicine's pragmatism (its willingness to use therapies that produce positive results, whether or not the therapies have been "proved" effective by scientific theory).

Holistic, New Age, and Folk Medicine

During the last few decades of the twentieth century, there was a surge of popular interest in therapies based on a wholly or partly religious interpretation of the healing process.

Holistic medicine. The holistic healing movement led this surge, though the precise meaning of "holistic medicine" varies from one healing system to the next. Among its meanings are an emphasis upon one or more of the following: "natural" therapies, patient education and responsibility, prevention, and the treatment of patients as "whole" people (that is, as more than just patients).

Common to all forms of holistic healing is the basic assumption that, as one handbook put it, "every human being is a unique, wholistic, interdependent relationship of body, mind, emotions, and spirit." The term "spirit" carries holistic healing beyond the scope of orthodox medicine's understanding of the relationship of body and mind in the healing process: the term suggests the existence of a deep belief that sickness and healing are the result of a complex, inseparable web of physical and nonphysical causes. Even when holistic practitioners urge reliance upon the body's own processes of growth and repair, their urgings are full of references to opening oneself to the inward flow of divine healing energy.

Persons who call themselves holistic health practitioners usually possess an outlook that is incompatible (that is, in total disagreement) with the framework of modern Western scientific medicine, which is founded on—and draws its conclusions from—the visible physical, natural world.

New Age medicine. The various religious and healing groups that make up the New Age movement endorse a holistic approach to health and medicine. They see every human being as a unique combination of body, mind, emotions, and spirit, and they hold the conviction that each person exists simultaneously in both a physical and a **metaphysical** reality.

New Age therapists use various physical and nonphysical tools and healing methods: crystals, so-called psychic healing, and a technique known as Therapeutic Touch, among others. These therapies seek to "channel" healing energies from the plane of metaphysical reality into the physical body. New Age crystal healing, for example, maintains that illness in the physical body is frequently caused by the loss of harmony in, or the disruption of, certain energies that exist in a portion of each person—specifically, the

Chinese herbal medicines are said to be effective, but they are neither regulated nor approved by the Food and Drug Administration (FDA). Some herbs can have fatal side effects if used improperly. Different combinations of herbs can be extremely dangerous. Some include natural **diuretics**, narcotic pain relievers, and poisons. But dangers exist with FDA-regulated drugs as well. Chinese herbal medicine is especially attractive to sufferers of chronic diseases who are offered little by Western medicine.

diuretic tending to increase urine flow

portion that extends beyond the physical world into the supernatural, or metaphysical, plane. Thus, this kind of healing seeks to achieve harmony between the physical and metaphysical parts of the person. New Age healers believe that crystals can receive and store energies that originate in the metaphysical plane.

Folk medicine. Folk and ethnic remedies continue to exist side by side with medical science. Botanical (plant-based) and herbal remedies are used both to promote better health and to cure illness without surgery or injections. Such remedies and a host of others—charms, prayers, rituals, incantations, massage, use of pressure points—are found in many forms, in many places, and among many peoples in the United States: the Pennsylvania Dutch, Mexican-American communities of the Southwest, and immigrants from the Caribbean, Southeast Asia, India, China, and Japan.

The ongoing presence of such folk and ethnic medical treatments may be a group's way of preserving its cultural identity. But it may also be evidence of exclusion from the mainstream medical system, a system that may have become too expensive for people to afford. There is the chance, too, that these remedies are the seeds that will produce a new, far less uniform medical system. Whatever the answer, society should respond thoughtfully and with great care to these and other alternative therapies, especially when its responses have legal and economic consequences.

The Challenge to Bioethics

The central ethical question raised by alternative therapies is whether they are genuine medical treatments or what doctors call quackery—deliberate attempts to deceive or defraud a trusting patient. It is reasonable to ask practitioners of alternative therapies to show that their therapies are beneficial to patients and to provide

▶ Traditional healers perform a healing dance at a health post in Nepal.

▶ A doctor of traditional medicine performs acupuncture on a patient.

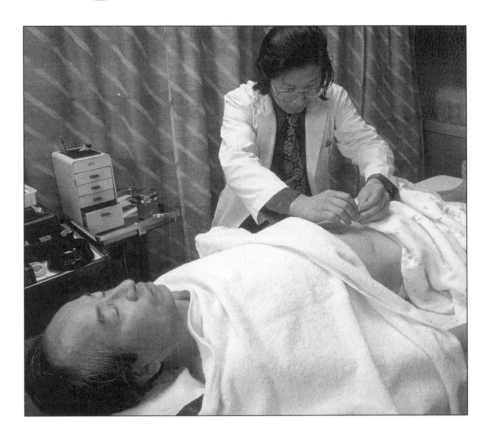

If an alternative therapy has worked over the centuries and is a successful part of an ethnic medical tradition, scientific Western medicine may have a great deal to learn from it. The fact that a traditional therapy has not been scientifically studied does not mean that it is invalid.

evidence that the methods they use actually cause the observed benefits. The concern of medical ethics should be to protect persons from harm, whether the harm is intended or not.

Although governmental agencies, health-care facilities, and insurance companies have limited resources for ensuring the welfare of the general public, they must all be prepared to make reasonable judgments about alternative medical practices that are based on systems of thought and belief at odds with modern Western science. Since quackery threatens both personal and public well-being, their judgments—though made and expressed with caution—must insist on a high standard of evidence for, and public scrutiny of, the healing claims made by alternative therapies.

Related Literature

Jack Coulehan's poem "Medicine Stone" (1991) describes a physician-narrator who picked up a medicine stone at Wounded Knee and participated in the sun dance ritual of the Navaho. When the doctor brings the medicine stone back into the city where he works, it loses its power. In the hospital, he says, the coyote is dead and the stone is of no account. Still he carries the stone, wrapped in sage, in a buckskin bag in his pocket, thinking that healing is not limited to Western medicine.

Alice Walker's story "Strong Horse Tea" (1968) portrays a poor African-American woman, Rannie, who cannot afford and has never used the conventional white person's health care. When her baby Snooks is dying from pneumonia and whooping cough, she desperately tries to send the white mailman to bring back a white doctor, but all the mailman does is summon a neighbor, Sarah, who wears magic charms sewn up in opossum skin and is known for her home remedies. Sarah knows the baby is dying and that nothing can save him.

BEHAVIOR MODIFICATION THERAPIES

Mental Health: Electroconvulsive Therapy

The fundamental ethical questions of behavior control remain constant: Who shall be controlled? By whom? How? By what right does one person control another, and within what limits?

Since the 1960s and 1970s, many developments have occurred in both the theory and the practice of behavior modification therapy.

Before 1970, behavior therapy was strongly criticized by some therapists as being too mechanical and rigid. It was believed, for example, that terms such as "behavior control" carried the message that some changes in behavior could be brought about by certain conditioning techniques. Critics thought, too, that these changes in behavior could be both involuntary and permanent. The behavior therapists of that era were also accused of attempting to impose certain goals on unwilling or unaware patients, and of using punishment and other techniques to bring them about. Critics saw behavior therapy as a way to impose the demands of the majority upon a helpless minority, such as prisoners, the developmentally disabled, and the mentally ill. Therefore, behavior therapists were viewed as instruments of those people in power who would not tolerate in others any deviation from what they considered right or normal.

Despite the protests of behavior therapists, brain surgery, shock therapy, and forced drug use were all lumped together as further examples of this harsh approach to changing behavior.

While a small percentage of early behavior-therapy practice did reflect these values, most behavior therapists shunned such methods of forced behavior change. Instead, they preferred a much fairer approach to setting goals and to changing behavior. Most behavior-therapy techniques lacked the power to bring about changes in behavior if a patient did not want them. In fact, most behavior therapists considered it unethical to "enforce" behavior changes against a person's wishes, even when the therapist believed that such changes would benefit the patient.

cognition the ability or process of acquiring knowledge

Cognitive Approaches in Behavior Therapy

In the early 1970s, behavior therapists began to explore the possibility of combining **cognition** and self-guided behavior change. Most cognitive behavior therapists assume that human behavior is guided in part by an internal "self." This self is made up of learned patterns of information processing that guide both immediate actions and general perceptions of the world. In turn, these perceptions have an important impact on emotions. Therefore, cognitive behavior therapists believe that, to change behavior, one must change the patterns through which information is processed. Ultimately, this change in information processing would lead to changes in a patient's behavior and emotional responses.

The 1960's rational emotive therapy (RET) of Albert Ellis was one of the earliest attempts at connecting both approaches to behavior therapy. Ellis's theory claims that emotional states, such as anger, happen as the result of an information-processing sequence in which an outside event triggers a set of beliefs that in turn trigger an emotional response. Effective treatment, therefore, enables the client to change irrational beliefs that lead to negative emotional states. This is done by directly challenging those irrational beliefs and by coming up with exercises to assist the client in learning that the beliefs are, in fact, incorrect. Take, for example, a patient who is fighting irrational feelings of shame and self-consciousness, which are based on an irrational fear of being made fun of. A rational emotive therapist might ask the patient to board a commuter train and loudly announce each stop to the other passengers. This activity helps the patient realize that such behavior, though appearing absurd and inappropriate, does not necessarily bring about public ridicule. However, even if other people were to make fun of such behavior, their responses would not be considered a catastrophe.

The most effective way of promoting both cognitive and behavioral changes is through performance-based treatments. In other words, the client actively participates in new behaviors that are very different from older, problematic ones.

Dr, Albert Ellis (1913–) is considered the grandfather of cognitive behavior therapy. He founded rational emotive therapy (RET), and was one of the architects of the sexual revolution.

Therapist–Client Relationships in Behavior Therapy

The ethical use of behavior-therapy techniques rests in part on the therapist's relationship to the client. It also rests on the therapist's sensitivity to social and political values. Particularly in cases where the application of a technique could cause pain, or where clients are powerless, ethical concerns still remain.

aversion procedure the application of unpleasant stimulation after an undesirable behavior has been performed

The use of **aversion procedures** has caused many people to criticize behavior therapists. When patients are given low-level electric shocks against their will, behavior therapists face a problem in which the benefit of the treatment has to be weighed against the client's rights. Even when aversion therapy seems to be the best treatment for stopping certain types of self-destructive behaviors, therapists are bound by their ethics to use other techniques to bring about change. Only when these techniques fail should aversion therapy be used. The use of aversion techniques with clients for whom fast, permanent behavior change is not absolutely necessary, or desired by the clients themselves, raises important ethical concerns.

The image of early behavior therapy among most therapists and the general public was extremely negative. Incorrect beliefs about the nature of behavior therapy were common. Behavior therapy was even put in the same category as brain surgery used to be to treat behavior problems. These misconceptions are less frequent now, when behavior therapy has become part of the mental-health mainstream.

Modern behavior therapy emphasizes client participation in setting goals. Clients' rights are weighed against society's values and expectations. Even in institutional settings, the application of techniques is much less forced. Behavior therapists are trained to apply their techniques only with strict safeguards for clients' rights.

> From the 1930s through the 1980s, people in the Communist USSR who dared criticize the Soviet state were categorized as "sluggish schizophrenics" by state psychiatrists and typically placed in mental institutions against their will.

Related Literature

Kurt Vonnegut's short story "Harrison Bergeron" (1961) takes place in the not-so-distant future when society has mandated that everyone must be equal. Nobody is allowed to be smarter or more athletic or more beautiful than anyone else, so everyone has to wear handicaps that equalize them to some common denominator. Smart people wear radios in their ears that blast loud noises at them periodically, making it difficult to think. Ballerinas wear weighted sacks that obstruct their movement; attractive people wear masks. When the genius Harrison Bergeron leads a revolt against this tyrannical equalizing, he is killed. The story makes us think about what social, economic, and psychological forces affect our own behavior and determine how we act.

Index to Volume 3

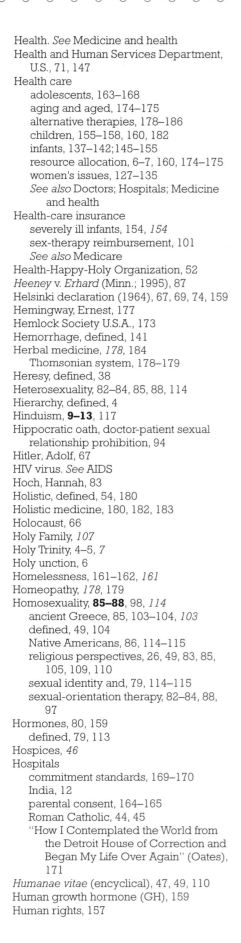